Full Unified Geometric Algebra
Amazing spatial arithmetic

gary-harper.com
Institute for Nagging Doubt
Dreamed up at
University of Hawaii
Beloved Alma Manoa

Index of Contents

Preface

Are points numbers? Yes!—they obey nearly all the rules of arithmetic except that point *a* "times" point *b* is the reverse of *b* times *a*: they are oppositely directed line segments. This kind of multiplication was called *extension* by its creator; aptly so because one point is *extended* to the other.

Its slight geometric enhancement of third-grade arithmetic generates an extraordinarily expressive **spatial** arithmetic whose simple rules can be understood by any child who knows the difference between left and right.

Such a child, when she enters high school, would be prepared to begin calculating with *well-dimensioned numbers* in physical space: scalars, points, vectors, bivectors, trivectors; whence she could easily imagine "imaginaries" as those numbers that perform some kind of spatial quarter-turn: their square is a half-turn, a direction reversal, multiplication by -1.

Forearmed with them, and ascending beyond physical space, she would be prepared, when she enters college, to express the diverse ideas of science *in a single pithy language*; a great advantage over her peers, who are now stuck with a rambling hodgepodge: scalars, complex numbers, vectors, matrices, tensors, linear algebra, non-linear algebra, Cartesian coordinates, homogeneous coordinates, barycentric coordinates, various geometries: projective, affine, analytic, hyperbolic ... on and on.

Spatial arithmetic encompasses them all, but it is not new, surprisingly: Hermann Grassmann came up with it in the early 1800's; and then late in that century William Clifford unified part of its syntax, which he called *geometric algebra*. At that time the full algebra was dimly understood, and even then by only a few, most notably Clifford himself, Peano and Whitehead. Now a few other luminaries have finally acquired much of it, about which they have become almost evangelic because ...

It has unprecedented simplicity and expressiveness. This will slowly become apparent via one *surprise* after another. Let us peek at a few to get started.

The first surprise is that addition of points generates their physical balance point, *complete with the correct weight there*. The second surprise is that subtraction of points—or more precisely, addition of two *separate-but-otherwise-exactly-opposite* points—packages them into an irreducible bundle, a *free vector*, v let's call it, *able to rove anywhere parallel to itself*. Its rovability owes to its weightless balance "at infinity".

Next surprise: extension of those same points generates an afore-mentioned directed line segment, a **bound** *vector*, **v**, *bound to the line thru those points*. Its bondage owes to extension's respect for summary (distribution over addition).

Extension filled-in that vector and gave it incremented dimension: **v** has bound dimension {**2**}; its *free part* v has free dimension {**1, 1**}-*without-magnitude*, better abbreviated to the well-known ordinary (free) *vector* dimension of {1}. Those dimensions arise because **points**, as the *scalable* *primitives*, have bound dimension {**1**}.

That may be almost too much surprise at the outset; so let us glance at a final one to see where we are headed: Retraction *undoes* extension, and such undoing loses whatever locus information extension had gained; so *retraction always produces a free result*, and of course it decrements dimension.

Retraction requires something *inside* to retract from (*projection*), which becomes *outside* (perpendicular) after the retraction. This is because any extension it had undone required something *outside* to extend to (*rejection*), which of course became *inside* (parallel) after the extension.

This complementary *outer–inner, perpendicular–parallel, doing–undoing* allows *free* extension and retraction to combine into a fully informative *geometric product*, more evocatively termed *extension-retraction*, more succinctly abbreviated to *extraction*.

That was Clifford's partial unification: a fusion of Grassmann's outer and inner products within the free sub-algebra. Our ultimate goal is to integrate that fusion into Grassmann's full geometric algebra, done in the *Synthesis* chapter.

Synthesis of the *free* algebra—but *only* the free algebra—is where most books on geometric algebra begin. If you studied one of them, you will be returning to your free foundations. If you hadn't quite understood them then, you may finally understand now. If you had understood then, I hope you at least enjoy integrating them into Grassmann's full algebra. I sure did.

For that journey, I must advise about an expositional flaw:

In my enthusiasms I sometimes just go on and on—I honestly don't know when to stop. So I occasionally suggest bypass routes for readers traveling fast who want just the essence. The present version, 1.04, provides a few more of those, some typo corrections, removal of one thinko and a few obfusco, and pervasive sentence-by-sentence attempts to enhance clarity.

(I love the correctability of ebooks and print-on-demand books—no author, except possibly Donald Knuth, ever gets a technical book right the first time, or even the nth time. Readers must always discard some garblge.)

My advice is to begin reading in sequence, flying low, until the algebra becomes utterly baffling, probably somewhere in the *Retracting* chapter. Then switch to the final encapsulating chapter, happily non-algebraic, to see if keeping going is worth your while.

I personally kept going thanks to the good ideas of Hermann Grassmann, William Clifford, Giuseppe Peano, Alfred Whitehead and all the others in the geometric-algebra community whose ideas delighted me so much. Of those alive now, I especially have to single out four: Lloyd Kannenberg for making Grassmann's and Peano's ideas available to us English speakers; David Hestenes, the evangelist of the free algebra who sparked my interest; John Browne, who has absorbed Grassmann's full algebra better than any other living mathematician; John Arthur whose deployment of the free geometric algebra in Electromagnetic Theory delights me. I have tried to replicate their good ideas, innovate on them, and discard whatever inferior ideas became evident, mine and theirs.

Of course, I must also thank my Sweetie and her maraming

mga anak for making good ideas so worthwhile, and innovation so much fun.

Whatever good ideas you may grasp herein, I hope you too can *R*eplicate them, *I*nnovate on them, and *D*iscard whatever inferior ideas that become evident, mine and yours; over and over again—*RIDding* I call it. **Creating** you probably call it. It's what Life has been doing for the past some 4,000,000,000 years on this planet. Grassmann finished his paltry 68-year burst of it a century and a half ago. Now is your time.

Adding Points

Addition is the most primitive calculation, so arithmetic always begins with it. *Ordinary* arithmetic begins with addition of natural numbers because they are the most primitive *scalars*—all other kinds arise from them. *Spatial* arithmetic augments ordinary addition with points, which are the most primitive *geometric numbers*—again, all other kinds arise from *them*.

Addition of points, as you shall now discover, generates their physical balance point, *complete with the correct weight there*, a mild surprise; ... and a big surprise: addition of *separate-but-otherwise-exactly-opposite* **bound** points generates a *free* vector, v let's say, able to roam anywhere parallel to itself. That is our main destination in this chapter.

After arriving, we shall explore our emergent freedom, and our foundational bondage. That exploration, I daresay, may induce some surprises about dimension and bases; unless you gain some independent *ah-ha!* thoughts along the way. So keep your eyes out, and your mind open.

The first step may not be a surprise: A sum is always of the same kind as its summands—apples plus apples are apples, not bananas. So you would naturally expect a point plus a point to be another point, not a line segment, right? But where would it reside? Would you guess the midpoint? You would be right, and addition's commutative rule enforces it:

$$a + b = b + a$$

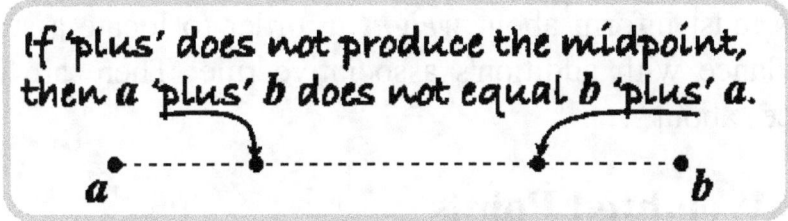

If 'plus' does not produce the midpoint,
then a 'plus' b does not equal b 'plus' a.

Commutativity requires a midpoint sum.

Here *a* 'plus' *b* mistakenly produces the point one fourth of the way from *a* to *b*. So of course *b* 'plus' *a* produces the point one fourth of the way from *b* to *a*. (Dashed lines indicate addition here, and hereafter.) This 'plus' does not obey the commutative rule—only a midpoint plus does. Knowing that, ponder addition's associative rule:

$$(a + b) + c \ = \ a + (b + c)$$

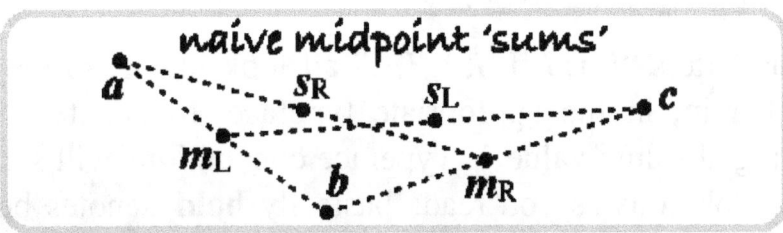

naive midpoint 'sums'

Naive midpoint sums invalidate the associate rule.

On the left, $a + b$ produces the Left *m*idpoint m_L. This is added to *c* to produce the final left midpoint sum s_L, as you see. On the right, $b + c$ produces the Right *m*idpoint m_R. This is added to *a* to produce the final right midpoint sum s_R. If the associative rule is valid they should be equal, but they aren't. What went wrong?

Insufficient information is what went wrong. A sum of simple points informs about *two* summands, so it needs *twice* the significance of either summand alone: a point plus a point produces 2 points. (An apple plus an apple produces 2 apples, right?) So *a* plus *b* should produce $2m$, rather than just *m* — points must inform about **weight** in order to locate a sum in accordance with addition's associative rule. There are three nuances about …

Weighted Points

First, weights convert *simple* points into *unit points*, more commonly known as **locations**: $a = 1a$. Second, the weight of a sum is the combined weight of its summands, as you might expect. Third, as you might not expect, non-unit points shall always explicitly display both *weight* and **location** like so: $2m$, *aa*, *bb*, $3s$, etc. This is in contrast to all higher bound elements, which encode magnitude and location in just one symbol: $\mathbf{v} = v\mathbf{i}$ for a bound vector for example, or $\boldsymbol{B} = b\boldsymbol{I}$ for a bound bivector, and so on.

(The letters "*i*, i, *I*, I, *i*, **i**, *I*, **I**" all look like 1, so they are used herein, in the appropriate typeface, to denote a *unit*, meaning absolute value 1. Typeface conventions will steadily become obvious as you read: basically **bold** denotes bound, subject to the convention that weightier typefaces denote higher elements.)

There is good reason for points to have explicit *weight****location** notation: they are unique among bound elements in lacking the spatial expanse needed to reverse their sign; so they have an intrinsic sign like scalars do. That sign is best viewed as part of their weight, not part of their location; so point weight and point location should be individually denoted.

Doing so has two benefits: First, it makes point locations units — *simple points*; second, it makes the sign of a point the

3

sign of its scalar weight. These careful distinctions enable us to see how weight information locates a sum:

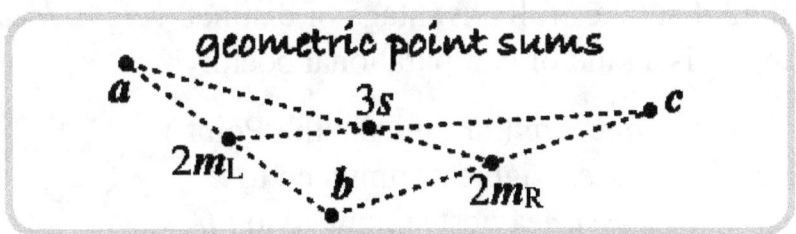

Weighted sums validate the associative rule.

The previous midpoints m_L and m_R have become $2m_L$ and $2m_R$. The two previous final sums s_L and s_R have become a single sum, $3s$, which lies closer to each of its heavier summands. How close exactly? That point is the *centroid* of the *a b c* triangle—its *center of balance*—whose distance from each unit summand is *twice* the distance from its corresponding weight-*two* summand.

(To see that, fold the triangle along a midpoint-line, and then explore the rise of the other midpoint-line above it.) This suggests that a generic purely positive sum *aa + bb* obeys this weight*distance rule:

$$\text{weight*distance } \boldsymbol{a} \;=\; \text{weight*distance } \boldsymbol{b}$$

Positive Points

For positive points, this rule expresses that the weight of *a* times its distance to the sum equals the weight of *b* times *its* distance to the sum. That puts the sum between its summands, closer to the heavier one, just as you would expect.

To validate the rule, be aware that it pays attention only to argument weight, not argument order, so it does not affect the commutative rule; which still mandates that the sum of

4

equally weighted points lies at their midpoint.

That fertile idea can be encoded into a recursive ***findSum*** algorithm that takes full advantage of distinct *weight*location* notation. It is a kind of computational poem:

> ***findSum***(aPoint, anotherPoint)
> ***set l***ightest summand to *ll*
> ***set h***eaviest summand to *hh*
> ***if*** $l = 0$ ***then return*** *hh*
> ***else if l*** and ***h*** are close enough
> ***then return*** $(l + h)h$
> ***else*** findSum($2l$ *midpoint(l, h), $(h - l)h$)

Each pass merely recomposes the sum as a doubly-weighted midpoint plus a residue point, $(h - l)h$, a ploy that appeals explicitly to the commutative and associative rules. Iterating it over and over generates midpoints of midpoints of midpoints … Their sums are equivalent to the original one, but their summands quickly close in around that sum. To illustrate, here are five iterations of findSum($5a, 2b$).

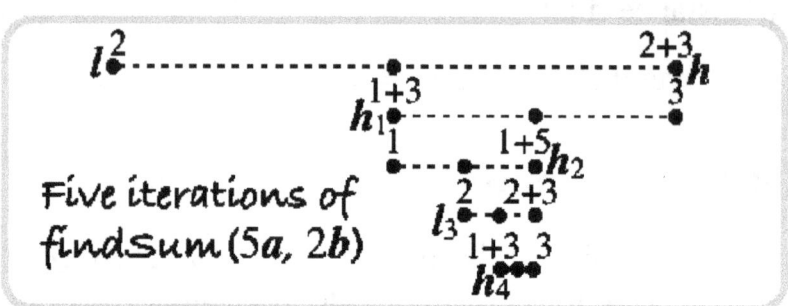

Five iterations of
findSum ($5a, 2b$)

Midpoints of midpoints validate the weight*distance rule.

See how quickly the summands are trapping the sum? They do so via three invariants for each pass. The first two are sum location and weight, obvious from the algebra: $ll + hh = ll + lh + (h - l)h$. The third is the weight*distance rule.

It is trivial within each pass because weights are equalized to produce a doubly-weighted midpoint. At each subsequent pass, the residue point is persistently propagated forward; which causes the weight*distance rule to be enforced across successive passes. Hence, it is valid all the way down by induction. Residue weight eventually either disappears, or else its location becomes indistinguishable from a midpoint.

For example, the first pass is $2l + 2h = 4h_1$ whose h-equalized weight is 2. The second pass is $3h_1 + 3h = 6h_2$, whose h_1-equalized weight is 3. These passes combine into $2l + 6h = 8h_2$ whose h-adjusted weight has become 6, as you may check.

That is the arithmetic, here is the spatiality: from midpoints of midpoints, it is clear that h_2 lies 3/4 of the way from l, and 1/4 of the way from h. That validates the weight-distance rule: 2*3/4 does indeed equal 6*1/4. Here you see elementary-school arithmetic collaborating with elementary-school geometry to enforce the weight*distance rule …

… without any measure of *distance*! The distance metric shall arise from self-retraction; so it must await the *Retracting* chapter. In the meantime, *findSum*() goes with the only thing available: the distance between its summands—whatever that may happen to be—which it successively halves. This means that, so far, *findSum*() has no *direct* way of knowing whether l_n and h_n are close enough. Fortunately there are two indirect ways:

Most obvious, successive midpoints converge quickly, so "close enough" could be redefined as "*if* n is high enough". For example, if findSum() reaches l_{30} and h_{30} without weight l having vanished, then those two locations would almost certainly be indistinguishable, within floating-point precision.

More subtly, there is no need to guess about distinguishability: if the point summands were expressed relative to a point basis, then their basis weights would distinguish them, even in the absence of a metric. Clearly, when l and h become expressed with basis weights that are indistinguishable, then they certainly are "close enough".

We will acquire such a basis shortly, but our journey thus far has already been remarkable. We started with arithmetic that arose historically from counting things like oranges and bananas. Who would ever imagine that such arithmetic might also be fruitfully applied to points?

Few have, in fact, and that is why we are only just now acquiring the following excellent idea: when the semantics of discrete points disciplines the arithmetic of discrete fruit, it unexpectedly generates the physical balance point, complete with the correct weight there. That is even true, and even richer, when it is applied to …

Negative Points

We do not know yet where the sum of two negative points might lie. In such a predicament, mathematics has a venerable stratagem: presume we *do* know, and then manipulate our presumption until we actually do. This idea arose several millennia ago with the ancient Greeks, but it was not really automated until Francios Viete did so in the late 1500's—he made the presumptions able to be *grammatically* manipulated.

As a specific example, we do not know where $-5a - 3b$ might reside; but we do know what its weight must be: -8 (sum weight is combined summand weight). So we cavalierly presume to know that $-5a - 3b = -8s$, even tho do not know where sum point s resides yet.

(This is the main maneuver in freshman algebra, where scalar x, rather than point s, is the unknown presumption. If history had been more appreciative, this ploy would now be called *Vietizing* the problem, or some-such.)

To find out where s is; we need to manipulate this presumptive equation; and the trick is always the same: apply a symmetry—a change that causes no change: change the equation without changing its equality.

We can negate each side of this equation—that doesn't affect its equality; but it changes it into a positive form: $5a + 3b = 8s$. Ah ha!—we know where s is *there* from the weight*distance rule: 3/8ths of the way from a and 5/8ths of the way from b.

That location must the same in the original equation because universal weight negations do not affect locations— that is one of the enlightenments of separate *weight*location* notation. Nor do they affect equality of the weight*distance rule either; so it remains as valid for purely negative summands as it was for purely positive ones: the sum still lies between such summands, closer to the heavier one (in absolute value). Here is a picture:

Both positive or both negative point sums.

8

Speaking of the weight*distance rule, there is a mysterious equation on the bottom of each figure, which looks just like that rule. Each is in fact a simple transformation of its top equation. For example, the top equation in the left figure becomes $5a + 3b = (5+3)s$ when the sum is separated into summand portions. If you balance this equation by leaving a on the left side, putting b on the right, each with their portion of the sum, you get $5(a–s) = 3(s–b)$, as shown. ¿What can this possibly mean?

It is in fact a precise algebraic rendering of the weight*distance rule: Each side has a weight, 5 or 3, multiplied by a point subtraction. Clearly, each subtraction informs somehow about the relation between the sum, s, and a summand, a or b. Why that relation would be summand-sum distance is a mystery we are about to resolve.

In anticipation, notice that it must be even richer than mere distance alone because, for example, $a–s$ and $s–a$ are negatives of each other—it must be some kind of *directed* distance. Indeed it is, as suggested by the arrows; but it is *even* richer than that.

Hint: ¿Would these two bottom equations be precisely true if their point subtractions were *roving* directed distances? Then their left side could be juxtaposed onto their right side to validate their equality.

To demystify such subtractions we must first rassle with summands having mixed signs, like $5a – 3b$ or its negation $3b – 5a$. Vietizing these sums generates $5a – 3b = 2s$ and $3b – 5a = –2s$, seen here. ¿Are there any symmetries we can apply to revert them to a purely positive form? Or else a purely negative one we just learned how to handle?

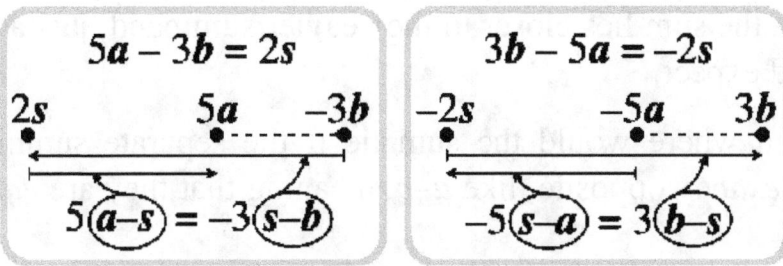

mixed positive and negative point sums

Yes. On the left in the equation $5a - 3b = 2s$ we can simply move b to the other side: $5a = 3b + 2s$. That produces a familiar positive form without changing equality; but it has the peculiar algebraic effect of exchanging the roles of summand a and sum s. The geometric effect is that original sum s lies outside its summands, a and b, on the line thru them.

Where exactly? A straightforward algebraic exercise shows that this sum is still properly located by the weight*distance rule, provided you give a distance that crosses one summand the opposite sign to a distance that doesn't. That is the geometric consequence of algebraic transfer across the equals sign. Tho straightforward and obvious, these algebraic manipulations are not geometrically enlightening.

Fortunately, there are non-obvious manipulations that are: transform each top equation directly into its corresponding weight*distance rule, as done previously. Give it a try—you will arrive at the bottom equations in each figure. Carefully notice that their directed distances are opposed to each other.

Which corroborates the need to give a distance that crosses one summand the opposite sign to a distance that doesn't. The geometric consequence is that summands with mixed signs generate a sum that lies on the line from the lighter summand (in absolute value) thru and beyond the heavier one. So, once again, the sum lies closer to the heavier summand, just as you would expect.

¿But where would the sum lie if the separate summands were *exactly* opposite, like *a–b*, meaning that they are *equally* heavy?

Free Vectors

The first thought is that weight here has become zero, meaning not that this sum *has* that weight, but that it *lacks* weight. Since zero times any point is zero, it seems that this sum must just be zero.

(Important and surprising aside: Since zero is commonly considered a real number, a person might suppose that unit-point subtractions "*have*" that weight. However, we are beginning to develop *distinctly dimensioned numbers*. Zero needs unique status among them as *no thing* with no properties whatsoever (*no value, no direction, no magnitude, no dimension*, etc.) for them to be able to vanish. So, in a well-dimensioned algebra, zero cannot be any kind of number, not even a real one; all of which do have properties. Consequently, when something (like weight) becomes zero, it has vanished, completely vanished—point subtractions truly do lack weight. As geometric numbers, however, they have formal context for it (coming up). **Zero lacks even that** and actually transcends numbers. It is especially useful anywhere diverse kinds of things might disappear such as set theory, vector spaces, programming languages, or even fruit companies. For details see *Fixing Nothing*.)

¿Is zero here really multiplying just "any point"? To find out, try approximating *a–b* as a limit. For example, start with *a–b*/2 and then successively halve the actual-minus-

11

appoximate weight like so: $a–3b/4$, $a–7b/8$, $a–15b/16$ … At each weight halving, the weight*distance rule scoots the approximate sum twice as far away along the line thru a and b. In the limit, the sum goes to infinity.

So the second thought is that this must be a classical "point at infinity", provisionally denoted as ∞_p. Which comes back to our question: ¿Does ∞_p lie on a's side, or on b's side? How could we ever decide?—a and b are equally heavy.

Maybe we can decide by examining how it was approximated. The dwindling sum was always closer to the heavier point a, the positive one in this case; so the answer seems clear: ∞_p lies on the line from the negative point thru and beyond the positive point. This seems natural and satisfying.

But consider this: we could have approximated on the positive point rather than the negative one. In that case the heavier point would have been $–b$; and the sum would have scooted off in the opposite direction, making it infinitely distant from itself! And that is just the beginning of its ambiguity: by approximating with different offsets, we can make this sum any distance from itself we wish. It seems to be just as ill defined and useless as 0 times ∞ (commonly written as 0/0).

It is in fact a geometric version of that uselessness, namely 0 times ∞_p. Consequently, the sum of $a–b$ *does not exist* within the algebra, meaning that its summands are *intrinsically composite*; or said more succinctly, they are *irreducible*. Fortunately such summands, as you shall now discover, are not only well defined, but are very *very* **very** useful.

Indeed, bundles of irreducible summands are amongst the most expressive novelties of geometric algebra; and this is our very first glimpse at them. How might they be used? For starters, the *a–b* bundle can translate a point: add it to *b* and you get *a*. Point *b* got translated from the negative point—its *tail* let's say—to the positive point—its *head*. ¿Can *a–b* similarly translate other points?

It could if it were able to move parallel to itself. Then you could place its tail over the point and add. *Poof!*—that new tail annihilates that point, leaving the new head as residue, a translation from tail to head.

In fact, the *a–b* summand bundle *is* able to move parallel to itself; and this is one of the most unexpected and amazing consequences of making points primitive. Speaking algebraically, if *a–b* truly is able to move parallel to itself, then there must be other points *c* and *d* such that *a–b = c–d*.

¿Can this equation be made purely positive, as done before? Sure: *a+d = b+c*. Now its left side is easy: sketch points *a* and *d* anywhere you want. Their sum lies at their midpoint with weight 2, call it *2m*. Sketch that point. It reveals a lacuna during conversion to positive points: whereas neither *a–b* nor *c–d* have a legitimate sum; both *a+d* and *b+c* do, which should be included: *a+d = 2m = b+c* .

Including that midpoint makes the right side, *b+c*, easy too: First sketch *b* anywhere you want. Its sum with *c* is also *2m*, which tells you exactly where *c* must lie: on the line from *b* thru *m*, twice as far away. Sketch that point. Here is a picture:

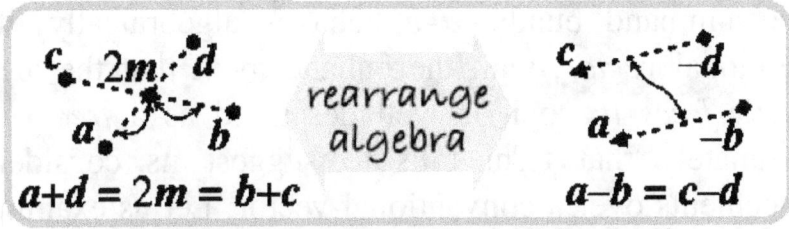

X-diagram translates a–b parallel to itself.

You just constructed an *X-diagram*, displayed on the left side of this figure. It is an exact geometric representation of the equation below it: dashed lines indicate addition, tiny curved arrows indicate equality.

It is composed of two congruent triangles, *a, m, b* and *d, m, c*; a *scissoring mechanism* that moves *a–b* to *c–d*, as shown on the right side. (When connecting positive–negative point-pairs by dashed-line addition, positive points are displayed as triangles, negative ones as diamonds). Congruence means not only that these summand bundles are parallel, but also that they have the same length-separation.

Length-separation is simply the *fixed distance* between the ends of a free vector. It causes all other free elements to also have fixed *separation* between irreducible ends. For a free bivector it is *area-separation*, for a free trivector it is *volume-separation*, and on up. The next chapter, *Extending*, will rassle with this.

If you can construct ten X-diagrams with arbitrary *a, b, d* points, correctly, without hesitation, then you truly do understand **algebra-derived** geometric freedom. Grassmann tried to explain it nearly two centuries ago, but he did not succeed; at least not with most mathematicians and nearly all pedestrians. If you hope to acquire his full geometric algebra, you will have to rassle with this freedom until you understand it in your bones.

The summand bundle *a–b* behaves algebraically like a conventional *vector*, with the enhancement that the algebra *explicitly frees it*; so it is well described as a *free vector*. Unfortunately that term fails to suggest its considerable enhancements over a conventional vector. Let us examine its conventionality first; and then start in on its enhancements. They will not become fully apparent until the end of the book.

Under summary, conventional vectors add with each other and scale, *period* (abstracted as a generic *vector space*). For free vectors, the adding is obvious: place the head of one on the tail of the other. *Poof!*—that tail annihilates that head leaving a head point subtracted from a tail point; another free vector.

Scaling is more subtle: $2(a–b)$ seems to generate $2a–2b$. Indeed it does, but when free vectors are deployed conventionally, they are constrained to be in *unitary form*, meaning that their head and tail have $1, -1$ weights.

In unitary form $2(a–b)$ becomes a free vector with twice the separation of *a–b*. This is easy to see simply by performing the sum $(a–b) + (a–b)$ tail-on-head: the tail again annihilates the head, leaving a free vector twice as "long".

More generally, any scaled free vector $s(a–b)$ can be converted into unitary form by finding unit points *c* and *d* such that $sa–sb = c–d$. This equation can be made purely positive just as before: $sa+d = sb+c$

This generates an X-diagram whose triangles are similar in the *scaling ratio*, rather than congruent. It causes scaling of a free vector to scale the distance between unitary endpoints, the canonical form of separation. Here is a picture for the case at hand:

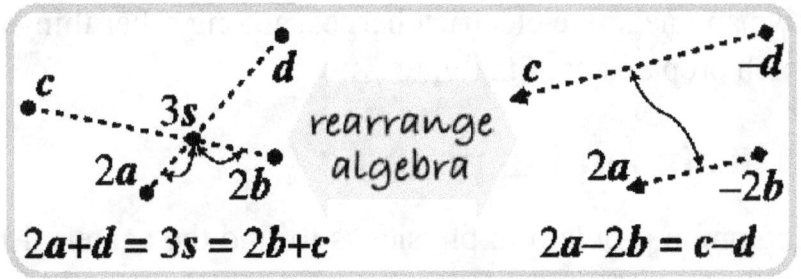

$$2a+d = 3s = 2b+c \qquad\qquad 2a-2b = c-d$$

Scaling a–b scales its separation.

Again, if you can easily construct ten such X-diagrams with $2a$, $2b$, d in arbitrary locations, then you have truly enlarged your understanding from mere free-vector rovability to its scaling. Particularly enlightening would be placing d somewhere on the line thru a and b, (maybe at a?). Give it a try.

That exhausts the conventional behavior of free vectors, so let us explore some enhancements. The first one was already mentioned: free vectors can translate points, something conventional vectors cannot do; at least not within the algebra.

But the idea is even richer than that: when combined with scaling, free vectors translate heavier points less far than lighter ones, a delightful and intuitive consequence, useful for physics.

For example, to translate $2p$ by $a–b$, first convert that free vector into an equivalent one with half the "length" and twice the head–tail weights (same formal separation, but not in unitary form). Its new 2-weight tail annihilates $2p$ leaving a new 2-weight head half as far away as $1p$ would have been translated.

A final delight is that negative points—since they annihilate positive head points—get translated in the opposite direction, from head to tail. This is useful for modeling

16

positive and negative electric charges, among other things. All of which prepares us, finally, to …

Begin adding points

A common problem in physics is to find the center of mass of a collection of point-like masses. The solution could hardly be simpler with the point algebra just developed: represent the masses as weighted points and add them together. The location of their sum gives the center of mass; and its weight gives the total mass there.

The way physicists actually do this could hardly be more complicated. They are still stuck using the conventional Gibbs-Heaviside vector algebra, whose vectors are (mis)interpreted as roving line segments. The line-segment idea was inherited from Rene Descartes, whose very first sentence in his 1637 *Geometry* reads "*Any problem in geometry can easily be reduced to such terms that a knowledge of the lengths of certain straight lines is sufficient for its construction.*"

This idea evolved into an algebra of *Cartesian coordinates* specifying orthonormal directed line segments. Since those lines have no formal locations, there is no direct way *within the algebra* to represent the fixed, point-like things that physicists often deal with. Such things have to be dealt with indirectly via an informal origin left *outside the algebra* to which "*position vectors*" are implicitly anchored.

Position vectors were independently concocted several times, but Maxwell's development is the one most revered by physicists. Renowned physicist Chandrasekhar asserted, on page 37 of his *Principia for the Common Reader*, that Maxwell's presentation is "*so completely in the spirit of*

17

[Newton's] Principia and illuminating by itself, that I reproduce the relevant sections (59–65, inclusive) of the chapter [IV], in their entirety." Here is the crucial part:

> We have seen that a vector represents the operation of carrying a tracing point from a given origin to a given point. Let us define a mass-vector as the operation of carrying a given mass from the origin to the given point. The direction of the mass-vector is the same as that of the [position] vector of the mass, but its magnitude is the product of the mass into the [position] vector of the mass.

Don't worry if you don't quite understand this on first encounter—physicists don't either. For example, Robert Romer, longtime editor of the *American Journal of Physics*, wrote …

> I can see in my mind the equation that (temporarily) drove me out of physics: … the position coordinate of the center of mass is given by
>
> $$x_{cm} = \Sigma_i^n m_i x_i / \Sigma_i^n m_i$$
>
> I had a sinking feeling. "This subject is just too abstract for me. I can never handle it." I closed the book. Obviously a new career plan was called for, even though I had somehow known for years that physics was what I was going to do, what I wanted to do with my life. It was not a happy moment.
>
> Fortunately for me, I was not prepared to give up so easily… Well, if $n = 1$, then $x_{cm} = x_1$. O.K. so far. Now what if we have two equal masses, $m_1 = m_2 = m$, say at $x_1 = 0$ and $x_2 = 10$. Yes, $x_{cm} = 5$, and that makes sense… And so, within a short time after returning to the book my career was back on track. I could do physics after all. I was beginning to understand [the equation], which was beginning to look a little less like an enemy and more like a friend. [65, p945]

It becomes a friend only after you learn its quirks, and tolerate them; but there is one more quirk that Romer mentioned only in passing. The position-vector recipe *only gives the location.* To get its mass there you must add all the masses together.

Happily, you had to do that anyway to get the location; so if you stash it away you need not do it twice. By contrast, a point algebra does everything for you automatically—position and mass—without you needing to explicitly spell out the details, as Romer just did for the simplest case, a line.

Position vectors are also used, in a similarly clumsy way, to model the *inverse-square* force laws of point-like particles. These forces arise when an expanding sphere of influence emanates from a point—such influence diminishes (it is *inverse*) in proportion to spherical area (a *square*).

Point algebra models these forces far better than position vectors; but we can only take an anticipatory peek now—full appreciation will not arrive until extension-retraction arrives in the *Synthesis* chapter.

Here is the peek: Point masses are inverse square under the force of gravity. Suppose we have two of them, *m*, *n*, whose position vectors are p, q. Then the conventional vector algebra gives the magnitude of the force as …

$$G \, mn \, / \, |p{-}q|^2$$

… where G is the constant of gravity and |p–q| is the length of the *ordinary* (not position) vector p–q, which is the distance between *m* and *n*. That distance is $\sqrt{(p{-}q)\bullet(p{-}q)}$, so this magnitude is sometimes written as…

$$G \, mn \, / \, (p{-}q)\bullet(p{-}q)$$

Notice that this is a function of four variables: *m*, *n*, p, q. It becomes somewhat simplified in conventional free geometric algebra as …

$$G \, mn \, / \, (p-q)^2$$

… because the dot product becomes incorporated into the geometric product via Clifford's unification. However, it is still a function of four variables. Point algebra reduces that to two, namely point masses *mp* and *nq*, where position vectors p and q have been converted into point locations *p* and *q*, appended by their weights there. The result is …

$$G \, mn \, / \, (\boldsymbol{p}-\boldsymbol{q})^2$$

This expression has several remarkable properties. Notice first the utility of distinct *weight****location** notation for points *mp* and *nq*—it allows weight and location to be used independently, but passed together in a single variable; like length and direction are passed together in a single vector variable. This is impossible with a position vector because an attempt to append mass would change its distance from the origin.

Most remarkable is that ***p–q*** is a *free vector* like p–q had been—the *same* free vector in fact. When passing weighted points *mp* and *nq* into the gravitational force function, we were working within the **bound** part of geometric algebra. And then, remarkably, the point subtraction in that function moved us into the *free* sub-algebra where extension-retraction works.

The utility of using point weight and location independently makes it crucial that we learn how to specify them independently. Let us begin with locations—*unit points*—weight one. There are two ways to give a generic point sum *pp+qq* that weight.

The obvious way is to simply coerce weight to 1 by setting *p* to be 1–*x*, say, and *q* to be *x*. That way, as we vary *x*, weights p and q will automatically sum to 1, so we will always generate unit points. The problem is to properly locate them.

To see exactly what kind of problem that can be, let us use this ploy to model locations on the so-called "real line". We have the sum $(1-x)p + xq$. Notice that when x is 0, this sum is p, and when x is 1 it is q. So, to better suggest their roles, let us rename p as o, the *origin*, because it looks like 0; and q as i, the *unit*, because it looks like 1. Here is a picture:

Varying x varies location indirectly.

As x varies, the location of $(1-x)o + xi$ changes indirectly and somewhat obscurely; but you should be able to convince yourself that setting x to e, for example, does generate a point that distance from the origin.

Despite its obscurity, the advantage of an explicit model like this is that it eliminates confusion students often have about the real line, that scalars are somehow "points" on it. They can't be because scalars and points have different dimensions — there is merely a *correspondence* between them.

But it is even more subtle than that: 0 is not even a scalar — it is *no thing*. Yet it too corresponds to a point on the "real line"; namely the origin, because that is what is left when unit point i has vanished. (Foreshadowing query: what *is* the dimension of scalars? points? free vectors? zero?)

The non-obvious way to model the real line is to stick with

the origin *o*, but convert unit point *i* into a unit vector, i = *i–o*. We do not lose *i*—it is *o*+i; but we do gain a direct and intuitive way to correspond scalar *x* with a point: simply use it to scale *how far* vector i moves *o*. Here is a picture:

modeled by point o, free vector i
$$o + x(i = \text{◆------▶})$$

$x = -1 \qquad 0 \qquad 1 \qquad\qquad e \quad \pi...$

Varying x translates the origin directly.

When *x* is 1, i moves *o* to where point *i* had been before it was tucked into vector i. When *x* is *e*, i moves *o* to the point corresponding to that scalar, and so on. When *x* vanishes, its product with i also vanishes, leaving *o* unmoved, the point corresponding to *no thing*. Such translation is far more intuitive than attempting to discipline point weights to sum to one; and simpler too.

This advantage is even greater in higher dimensions. Suppose, for example, that we are trying to locate points in physical space using four non-coplanar points *o*, *p*, *q*, *r*. Making them sum to weight one is not too hard, but locating that sum while doing so would be almost impossibly obscure for most people, tho an algorithm could do it easily.

By contrast, if we set p = *p–o*, q = *q–o*, r = *r–o*, then locating is almost trivial: simply translate unit point *o* along p a certain distance, then along q another distance, and finally along r. This is the essence of familiar Cartesian coordinates, but now **enhanced with an explicit origin o**.

Another advantage of this approach, beyond its familiarity, is that once you have located the point, you may give it any *weight*, *w*, you want, expressed like so: *w*(*o*+pp+qq+rr). This

22

is the *locate-then-weight* strategy. There are two others.

First, using the original o, p, q, r points, you could simultaneously locate and weight their sum. This compounds the obscurity of merely locating, making this approach virtually incomprehensible for most people.

Second, you could *weight-then-translate* like so: ($wo+pp+qq+rr$). I had called this "*weight-then-**locate***" in the initial version of this book, but that is a little misleading because it locates only indirectly: vectors p, q, r here move wo only $1/w$ the distance they would move unit point o.

This was demonstrated geometrically already; but it is easy to demonstrate algebraically: First, express it more succinctly as a weighted point plus a vector: $wo + v$, where v = $pp+qq+rr$. Then, compare this expression with the *locate-then-weight* strategy in the same form: $w(o+v) = wo + wv$.

Both strategies end up translating a weighted origin wo. The *locate-then-weight* strategy translates it to exactly the same place it would have translated an unweighted origin o. The *weight-then-translate* strategy translates it only $1/w$ that far, as you see.

That is a serious complication for a human; especially a human accustomed to directly locating things with Cartesian coordinates. So we shall stick with the *locate-then-weight* strategy hereafter for specifying weighted points. Doing this well requires the …

Dimension and basis of primitives

One of the most remarkable consequences of Grassmann's extension is that *for the first time in history dimension became well defined as a formality **arising within** the* *algebra*. This generates distinctions that either had not been made before, or else had been left *outside* in the interpretation.

The most important distinction is the formal one between *bound spaces* and *free spaces*, induced by a basis whose scalable primitives are either *bound* or *free*, respectively. To define those spaces, we first need to formalize primitive …

Dimension

Dimension is always based on a primitive foundation. Here is how Grassmann presented it in the first half of his first book (but more esoterically):

- ***Extension is the generator of dimension.***
- ***Extension with a primitive increments dimension.***
- ***The primitives in geometry are points.***

The first two ideas are familiar from conventional free geometric algebra, whose scalable primitives are free vectors, dimension one, misinterpreted as roving line segments.

The line-segment idea was implicit in Euclid over two millennia ago; subsequently frozen in by Descartes nearly four centuries ago; then made into an *imaginary* algebra by Hamilton not quite two centuries ago; and finally dismantled into our conventional *non-imaginary* vector algebra more than a century ago by Gibbs and Heaviside. It has been a wild ride.

¿Wouldn't it be more natural—*since all geometric figures are composed of points*—to make points the scalable primitives, as Grassmann did? If he is right, then—*since extension with a primitive increments dimension*—points must have dimension one, like free vectors do.

As you may have guessed by now, this is not a contradiction: addition of *separate-but-otherwise-exactly-opposite* points generates a free vector, which we now know is an intrinsically composite bundle of points, not a line segment. Intrinsic separation is the distinction we need to formalize. For that we first need to formalize the dimension of what is being separated, namely points.

We already denote bound things in **bold**, so bound dimensions shall also be denoted that way: points shall have bold primitive dimension **1**. Which means that free vectors have bold, intrinsically composite dimension **1, 1**.

This kind of composition arises from *separate-but-otherwise-exactly-opposite* bound summands, whose sum lacks magnitude. There are other kinds of composition, described in the next chapter, arising from incompatible dimensions or insufficiently intermingled spaces.

To denote all such composition in a uniform way, braces shall henceforth enclose their dimensions. This has the benefit of providing an explicit context for dimension, crucial for denying zero any such context.

Under this enclosing discipline, points have dimension {**1**}, but free vectors have dimension {**1, 1**}-*without-magnitude*. *Without magnitude* is the algebraic formality that distinguishes free vectors from fixed points; which do have magnitude, namely their non-zero weight. *Without magnitude* does not require a spatial metric, so free dimension is well defined even in the absence of a metric.

However, it is excessively clumsy as denoted so far. If we were to abbreviate {**1, 1**}-*without-magnitude* as {*1*} using regular italic font, we would make free-vector dimension consistent with conventional-vector notation, while still maintaining the crucial distinction between bound and free, like so: points have bound dimension {**1**}, free vectors have free dimension {*1*}.

These distinct dimensions each have the numeric value needed for primitives in a basis; a fruitful idea that induces a formal distinction between bound spaces and free ones.

In preparation for that distinction, ponder the other fundamental dimensions, that of scalars and zero. Scalars have dimension {0} (not points, as Euclid implied), meaning that scalars lack dimension (or even geometric locus, which points do have). A common way this dimension arises is from retraction (dot product) of a free vector with another one, which produces a scalar having dimension {*1–1*} = {0}.

Zero itself has dimension {}, meaning *no context for dimension*; which allows things to vanish without leaving a dimensional residue (such as a so-called "zero vector", a "real zero", a "complex zero" … etc.—oxymorons all).

Basis

We have already seen that four non-coplanar points $o, p, q,$ r can generate points anywhere in physical space. The syntactic requirement for them to be a *basis* is that they be *independent*, meaning that none of them can be generated by scaled summary of the others. That *algebraic* discipline keeps them *geometrically* non-coplanar.

This is easy to understand incrementally: Two points can generate a line of points under addition if they are not co-located, meaning that one is not a scaled version of the other.

26

Then, to generate a plane of points, we need a non-collinear point off that line, meaning that it is not a scaled summary of the first two points. And to generate a volume, we need a non-coplanar point off that plane, not a scaled summary of the first three points. And so on.

The o, p, q, r basis generates a *bound four-space* because its four basis primitives are bound, each with dimension {**1**}. Bound spaces actually only need one point to bind them—the *origin*—as already alluded to, so it is useful to further distinguish them by the *coordinates* used to scale the basis elements. Here, they are pure *point coordinates* obviously.

Point coordinates are nothing more than so-called *"barycentric coordinates"* stripped of their simplex length-area-volume baggage. They were developed by August Mobius in his 1827 *Barycentric Calculus*. The reason they acquired their baggage is that Mobius, in his preliminaries, declared that individual points are simply abbreviations for ends of line segments.

There is an historical reason for using line segments to define points: line segments are scalable and points are not —or so everyone thought because Euclid had said so. However, as you have seen, points actually need to be scalable to obey the rules of addition; and Mobius gave them that scalability via the scalability of line segments.

Grassmann used the same ploy in his first book, independently, for the same scaling reasons.[p154-162] However, his second book, published eighteen years later, applied *"the general concept of addition … specifically to points"* rather than developing points (as he had in his first book) as *"abbreviated notations, as Mobius would have it."*[p129-131].

Grassmann's point coordinates become *anchored free-as-possible coordinates* when they are applied to the o, p, q, r basis that o, p, q, r transforms into. Such a basis generates the same bound four-space; and its primitives are just as independent—subtraction of the origin o from a basis point p or q or r does not enable the other basis elements to generate that point.

Such anchored coordinates are nothing more than so-called "*homogeneous coordinates*" augmented with formal free–bound distinctions, which strips them of their uniform misinterpretation in terms of line segments.

It was Mobius again who gave them that interpretation in order to get the rules of addition to work on his origin. Mobius, in effect, *artificially homogenized* his coordinates by making the weight of the origin scale a line segment, like his three other coordinates did.

Homogeneous coordinates are used in computer graphics to locate and weight points; but their historically unfortunate artificial homogeneity prevents programmers from transparently distinguishing between points and free vectors.

This makes the intrinsically simple locate-then-weight operation needlessly confusing—the common graphics practice of scaling a free vector to weight a point is truly perplexing. However, history demonstrates that clever people are able to accommodate even the most bizarre perplexity if it is expressive or appealing—witness imaginary numbers or the virgin birth.

There is a subtlety about weighted points in a free-as-possible basis: They are represented in the basis as a weighted origin plus a free vector translator, $wo+v$; but the *translation vector* v does not directly determine the point's *location* because it moves the origin only $1/w$ as far as it would have moved the unit origin.

This has already been mentioned, but it needs to be emphasized because the **actual location** of $w\boldsymbol{o}+\mathrm{v}$ is $\boldsymbol{o}+\mathrm{v}/w$. It is a unit point whose *location vector* is v/w. To check this, simply multiply this unit point by w, and you get $w\boldsymbol{o}+\mathrm{v}$, the basis representation of its associated weighted point.

Hence, when adding weighted points to find center of mass, you must carefully distinguish between the final *translation vector* v and its associated *location vector* v/w. Notice in particular that if weight is negative, the translation vector and its location vector *point in opposite directions*; a disconcerting thought that should motivate carefulness.

In summary, a bound-as-possible basis exposes free vectors as roving irreducible sums of points. This kind of basis is transparent about their separate-but-opposite freedom, but clumsy about articulating it. Conversely, a free-as-possible basis incorporates that freedom right in the foundations, and is expressive about articulating it, but completely opaque about its intrinsic composition.

Perhaps the most important advantage of a free-as-possible basis is that it engages retraction right at the foundations too. Retraction, as you shall eventually discover, can articulate perpendicularity; and can also articulate a spatial metric.

However, retraction only works in the free sub-algebra because it undoes extension, so it loses the locus information that extension had gained. (Extension, as you shall shortly discover, works in the full algebra, bound and free, where it can articulate parallelness.)

The basis for the free sub-algebra is nothing more than the purely free part of an anchored basis, meaning the free vectors within it.

(So, such a basis is intrinsically homogeneous, containing only dimension *{1}*. And a pure point basis is also intrinsically homogeneous, containing only dimension {**1**}. Ironically, only a so-called "homogeneous" basis is intrinsically non-homogeneous, containing dimensions {**1**} and *{1}*. History often saddles us with inappropriate terminology that we—for kindness to our intellectual progeny —really ought to revise for clarity. "Homogeneous basis" would become *anchored free-as-possible basis*, or simply *anchored basis*.)

For the physical-space case at hand, the free vectors in an anchored basis generate a formal *free 3-space*. This basis engages retraction, so it too can have a spatial metric.

In summary, here are all the spatial distinctions generated by points, dimension {**1**}, and free vectors, dimension *{1}*:

An ***n**-point basis* generates a *bound **n**-space*. That same space can also be generated by its corresponding *anchored free-as-possible **n**-basis*, which contains just one point, the origin. This encoding of the bound space can engage a spatial metric. The ***n*** in a bound space, and in both of those bases, specifies the number of independent points needed to generate that space (directly or indirectly). Those ***n*** points generate a *spatial expanse* of ***n**-1.

That expanse arises from the $n-1$ free vectors in the anchored basis, which themselves constitute an *($n-1$)-vector basis*. They induce a *free ($n-1$)-space*; and it may also engage

30

a spatial metric. The $n-1$ in that basis, and that space, specifies the number of independent roving separations needed to generate that spatial expanse; the elbow room generated by the corresponding **n**-point basis.

Spatial expanse has been considered the only valid geometric dimension for millennia. The idea was frozen in by Euclid, who explicitly denied points magnitude because they lack spatial expanse. That blunder denied points the ability to generate higher spaces, for which their weight is as crucial as the length of line segments is.

Lack of weight and lack of dimension have persistently inhibited points from participating in an algebra as bona-fide numbers. Only recently have points been permitted weight and dimension; but those ideas have been confronted with considerable intellectual inertia against them. That inertia is *overwhelming* against the idea that points are numbers.

Until it dissipates we will be unable to acquire the rich geometric dimensions generated by the full geometric algebra. To make some of these distinctions, there have been various ad hoc …

Historical attempts

After Gibbs and Heaviside developed conventional vector algebra, geometers realized that it handles points poorly. Lacking knowledge of Grassmann's point algebra, their early solution was to introduce a point outside the algebra as the origin, as with position vectors.

It was soon realized that leaving the origin in the interpretation is insufficiently informative; so points were introduced formally. Early attempts did this with logical axioms; but they have been abandoned as clumsy and impractical for computation—logic calculates poorly.

What calculates well is arithmetic of scalar coordinates—Descartes' revered innovation—so points have recently been defined as special sets of coordinate tuples obeying certain rules. To illustrate, here is the definition of a popular point space from a 2011 book on geometric methods for scientists, adapted to notation in this chapter.

> An *affine space* is a triple $< P, V, + >$ consisting of a set P (of *points*), a vector space V (of *translations*, or *free vectors*), and an action $+: P \times V \to V$, satisfying the following conditions:
>
> A1 $p+0 = p$ for p in P
> A2 $(p+u) + v = p + (u+v)$ for p in P; u, v in V
> A3   For p, q in P, there is a unique v in V such that $p + v = q$.
>
> The unique v in rule 3 is denoted as $p \to q$, or sometimes **pq**, or even $q\text{–}p$.

Do you see what is happening here? The point algebra that *naturally emerges* when points are treated as numbers is here being *artificially imposed*. No understanding is evident that points P automatically generate vectors V under addition; or that affine rules A1, A2, A3 are automatic consequences of elementary-school arithmetic applied to points.

Especially lacking in understanding is A3, which artificially imposes the rule that vectors translate points like so: $p + v = q$. This rule automatically arises from the arithmetic of points when it is understood that a "*translation*" v moves point p via endpoint cancellation; meaning that v is nothing more than a roving, intrinsically composite bundle containing a head point and a tail point.

Failure to understand that is here corroborated by the

mention that v might be denoted as "*even* q–p"; which shows that such notation was not arrived at by the *geometric semantics* of point addition, as developed in this chapter; but rather by *algebraic syntax*:   point p was simply moved to the right side of $p + v = q$; the idea being that …

> *Well, the algebra does generate* v = q–p *after all; so we can even use that notation for vectors, rather than $p$$^{\rightarrow}$$q$, or pq, tho it does seem strange.*

Cavalierly moving a point across the equals sign does not provide even the slightest clue that it suddenly induces a vanished sum at infinity, which is just as ambiguous and useless in an algebra as 0/0 is. Failure to understand that only its summands at finity are useful is clear from the definition of another popular point space, *projective space*.

Projective space in that same book is defined on the plane in a traditional way by *artificially appending* to the set of points P at finity a further set of points P_∞ "*at infinity*". The motivation for these points is appealing: they make parallel lines intersect far away, as they seem to, thereby making all lines in the plane intersect.

This may seem to be a wonderful simplification; but in fact it immediately and unknowingly cripples this space by denying points full expression within an algebra.

Enlightening exercise: try to use addition of points aa and bb at finity to generate a point p_∞ "*at infinity*". Please try. You will discover that you must make weights a and b infinite, but with a difference of 1, an impossibility.

Fortunately, *parallel lines already intersect in the full geometric algebra*. Let's rassle with this: two non-parallel lines on a plane *share* a unit point i; meaning that each line is

33

generated by scaled addition of *i* with some other point on that line. Similarly, two parallel lines *share* a unit free vector i; meaning that each line is generated by scaled addition of i with some point on that line—*that free vector is a bona-fide* **algebraic intersection** *of those parallel lines*.

Think about it: i's parallel rovability at finity really does accurately represent a conventional so-called *"point at infinity"*.

This means that a projective space is already included in any bound space, just as an affine space is; so both spaces can be directly articulated by spatial arithmetic at finity; a huge improvement over the current various ad hoc machinations.

The historical difference between the arithmetic of those spaces is how they are restricted and what is ignored. In a nutshell, a projective space ignores scaling the separation of a free vector, effectively working with unit vectors. An affine space pays attention to that scaling, but ignores scaling the weight of a point, effectively working with unit points, just as a projective space also does. Both spaces ignore the spatial metric.

There is nothing sacrosanct about these particular ignorances—they are just properties that some mathematicians wanted historically. You may want different ignorances.

How the ignorance is implemented for an affine space can be seen by going back and scrutinizing its rules. Notice that they only allow points to add with vectors—points are not allowed to add with other points, a ploy that implicitly denies them non-unit weight. (A ploy that also precludes understanding that free vectors are bundles of points.) A

34

metric is just ignored as one of those appurtenances irrelevant to an affine space.

¿Does it make sense to install such ignorance in ad-hoc definitions that unknowingly cripple points as numbers? Wouldn't it be better to knowingly use points as numbers, but discipline their use?

For example, if you want to articulate only unit points, just translate the unit origin with free vectors. If you want only unit free vectors, just install them in the basis and make sure you don't scale them; or if you must, normalize them afterward. If you want to ignore a metric, just don't appeal to it.

The best feature of this ploy is that it may happen that you actually *need* weighted points, or scaled vectors, or a metric; and it is comforting to know that they are available. This is a common scenario in computer graphics.

The responses to this appeal that I have heard claim that ad hoc definitions capture important features that are not present in geometric algebra. A projective space, for example, requires the concept of *duals*.

Actually, duals were present in geometric algebra right from the beginning, even before their concept became widely known in projective contexts. Grassmann knew nothing about projective duals (indeed, nothing about projective spaces or even affine spaces) but he intuitively implemented duals as reflections from the ceiling.

Only recently has John Browne made such ideas computationally tractable. Browne uses ceiling-reflection to make his extensions and retraction pay *explicit* attention to the embedding space, just as projective geometry does. (By contrast, the extension and retraction developed in the book you are reading pays only *implicit* attention to the embedding space, via the basis.) If you want to see how elegantly *Grassmann Algebra* can articulate projective geometry, study Browne's book by that name.

The other popular space often implemented via an ad hoc definition is *Euclidean space*, the most popular space of all. Here is the prelude to its definition:

> A *Euclidean space* is a real vector space V equipped with a symmetric bilinear from •: V×V → $\mathbb{R}$ that is *positive definite*. More explicitly , • satisfies the following axioms: …

Then are listed six rules that • obeys. They *artificially impose* the rules that *naturally emerge* when retraction is developed as the undoing of extension, and given a metric. In particular, such undoing requires self-retraction to produce a positive scalar if it doesn't vanish, which is the meaning of *positive definite*.

A Euclidean space, translated into geometric-algebra terms, is a free space in which retraction is equipped with an orthonormal metric. There are no points in it, like there were in the other two spaces; so it is impoverished despite its popularity, or maybe because of it.

There are no bound vectors in it either, or bound bivectors, or free ones … on and on. Nor were there any in the other two spaces. Every one of the just-defined spaces is impoverished relative to the full geometric algebra. When we actually need their poverty, why don't we just hobble geometric algebra to provide it? Why are we still using ad hoc, poorly comprehended definitions going on several thousand years now?

Intellectual inertia.

Extending

Extend one point, **b**, to another, **a** (whoa: ¿**b** to **a**?), and you finally arrive at a bona-fide directed line segment; a ***bound vector***, **v** let's denote it in **bold**, whose bondage will shortly be derived.

This line segment has well-defined bound dimension {**2**} because a point's extension with another point increments its primitive dimension of {**1**}. Before point-extension arrived, line segments were un-derived, tacitly primitive, and ill-dimensioned for some three thousand years; one of the great calamities of mathematics.

Bound vectors

Before scrutinizing them, we must become clear about the geometric order of their arguments, **a** and **b**. ¿Do we want **a** to be the tail and **b** the head, as seems natural? Which is to say, do we want to *extend **a** to **b*** as the *wedge product* ∧ would, were it applied to points? ¿As Grassmann did in his various notations applied to points?

No, we want our new bold vector **v** to have the head–tail order of our old regular vector v = **a–b** because these two kinds of vector have a symbiotic relation, next up. We certainly don't want them heading in opposite directions—that would be needlessly confusing. Therefore, we shall *extend **a*** *from **b***, symbolized as **a‹b**. Such *explicitly directed*

38

symbolism for a *directed* operation like extension is a further advantage over the un-directed wedge ∧. Here is a picture:

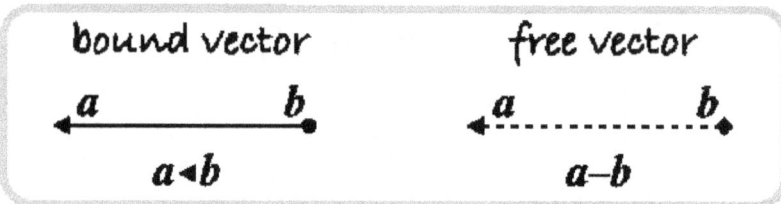

Symbiotic vectors, bound and free.

The first thing to notice about these vectors is that both $a\!\cdot\!b$ ("*a* extended from *b*") and $a\!-\!b$ ("*a* minus *b*") *neg-commute* because exchanging their arguments reverses their direction:

swapped arguments negate each vector

$$b\!\cdot\!a \;=\; -(a\!\cdot\!b) \qquad b\!-\!a \;=\; -(a\!-\!b)$$

Seminal neg-commuting rules, bound and free.

(Neg-commuting is usually termed "*anti-symmetric*", a term that is obscure and improperly specializes the idea of symmetry; or sometimes termed "*anti-communting*", which is not as pompous but still needlessly obscure. *Neg-commuting* can be understood by any smart child in third grade, my benchmark for mathematical exposition.)

Extension's neg-commuting automatically extends to bundles of points like u and v: v⁴u = –(u⁴v), as may be checked *algebraically* by descending to points. (*Geometrically*, what u⁴v or v⁴u mean is a crucial task of this chapter—it is not what you may think if you are familiar with conventional geometric algebra.) These simple, easy-to-comprehend rules have intricate, hard-to-comprehend consequences, as shall become apparent.

For example, the second thing to notice is that free vector v = a–b extended from b produces bound vector $\mathbf{v}$ = $a \triangleleft b$: Extension's neg-commuting makes $b \triangleleft b$ vanish because it equals –$b \triangleleft b$. (Its vanishing is geometrically obvious: a point extended with itself fails to produce a line segment.) This relationship between free v = a–b and bound $\mathbf{v}$ = $a \triangleleft b$ is so crucial that it has special terminology: free v is the *free part* of bound $\mathbf{v}$.

(Vista ahead: One reason a free part is crucial is that its extraction is the most fundamental kind of retraction—it decrements numeric dimension. Retraction itself is crucial because self-retraction of course descends to dimension $\{n$–$n\}$ = $\{0\}$, namely a scalar, so it establishes the metric; but it requires free arguments. This means that the seminal metric is *free separation*. Then, **bound magnitude** derives from that as separation of a free part. This is obvious here: the **length** of bound $\mathbf{v}$ clearly equals the *length-separation* of free v because they have the same ends. Similarly, *same ends* shall make the equivalence of **area** and *area-separation*, **volume** and *volume-separation* equally obvious. And on up.)

Bound $\mathbf{v}$ was generated by placing the tail of its free part on a point and then extending, which fills in free v from tail to head. This is the *tail-poof* method of vector extension, similar to the *tail-poof* method of vector addition. Both are potent simplifications that shall be generalized and used again and again. (Are you beginning to see why it is crucial to have free vectors and bound ones expressed in the same order?)

Enlightening exercise: Calculate free v extended from a, rather than from b. This is important. The neg-commuting of *subtraction* will counteract the neg-commuting of *extension* to almost magically produce the *same* result: $v \triangleleft a = v = v \triangleleft b$. Please try.

(I am doing my best to help you understand, but you need to keep your pencil handy—pencil-deprivation is the primary cause of incomprehension.)

This gives a preliminary peek at the geometric intricacy that algebraic neg-commuting can induce. It means that free v extended from *any* unit point $c = aa+bb$ on the line thru *a* and *b* will generate $a \triangleleft b$:

$$v \triangleleft c = v \triangleleft (aa+bb) = av + bv = (a+b)v = 1v = a \triangleleft b$$

Please don't skip that elementary-school algebra—it's almost poetry. It means that directed line segment **v** is free to move anywhere along the line thru itself. Which raises the question: ¿Is bold **v** actually free to move *anywhere* parallel to itself, like regular v is? Not just along that line?

That seems reasonable—it is the current universal presumption in the conventional geometric algebra. The rovability of this line segment, in fact, initially seemed so obvious to Grassmann that he declared "*a new definition is warranted, [αβ] and* [parallel-translated] *[α'β'] are assumed to be equal.*"

However he decided, in his very next sentence, to "*ascertain the extent of that warrant.*" He then proved it is true by assuming it is!—see for yourself on page 52 in his first book. That underscores the tremendous intellectual inertia that roving line segments have had for centuries, thanks largely to Descartes.

Grassmann eventually discovered that line segments derived from point extension can only move along the line

41

thru themselves. This is trivial to demonstrate—reverse the previous algebra: If unit point c does not equal $aa+bb$ then $(a-b)\triangleleft c$ does not equal $a\triangleleft b$—there will be intrinsically non-a-or-b points that cannot be eliminated; give it a try. Said geometrically, no extension with a point outside bold **v**'s confining line can reproduce it—that vector is "***bound***" to that line, as Grassmann declared belatedly in the last half of his book.

Such bondage gives any bound vector **v** an invariant *directed-line* and invariant *length*. Its directed-line is its oriented confining line; its length arises within that line, like so: head and tail points translated differently there do not reproduce **v** under extension. You may check this using free-vector translator $a-b$, differently-scaled. Owing to these invariants, it is often useful to explicitly decompose **v** into its scalar *magnitude* times its unit ***directed-locality***:

$$\mathbf{v} = v\mathbf{i}$$

Similarly, every other *bound element* can be expressed as a magnitude times a directed-locality. A *bound element* is simply a point or an (effective) extension of them; so it is intrinsically atomic as a sum. (Peek ahead: $\mathbf{v}\triangleleft a$ is an *effective* extension of points because you can express v's tailpoint at a, which leaves only v's headpoint in the extension.)

By convention, magnitude and directed-locality are both made positive, by spatial reversal if necessary. This convention must be violated for points because they do not have the spatial expanse needed to remove a negative sign; so signed-*weight****location** notation is usually preferable for them.

However, *magnitude****directed-locality** notation is

42

sometimes conceptually useful for points because it would extend the idea of *orientation* to them. In essence, this simply transfers the sign of a point's weight to its unit-point location. We shall ponder such universal orientation after all physical manifestations of it have been acquired.

Adding bound vectors

Adding free vectors was conceptually simple because they themselves are an addition, easily juxtaposed by parallel transport. Adding bound vectors is more difficult because they arise by extension of points; a product that can only be juxtaposed by sliding along a confining line; and which is not addition, but merely has a familiar relation with it:

$$(a+b) \blacktriangleleft c = a \blacktriangleleft c + b \blacktriangleleft c$$

(I silently presumed this rule a page ago.) When applied to scalar multiplication, this is one of the fundamental rules of ordinary arithmetic, like addition's commutative and associative rules. It's justification for point extension is that it corresponds to extension's geometric interpretation.

Best of all, it extends to higher dimensions all of the free-and-bound magic already induced by the rules of addition. In other words, this new rule works as beautifully for point *extensions* as the previous rules did for points themselves.

It is of course inherited by bundles of points like a, b, c; and the order of its extension arguments can be reversed by neg-commuting them; as you might have fun trying. Historically, it was labeled the *distributive* rule because one product gets distributed into two, reading from left to right across the equals sign.

We have the opposite goal: we want to *collect* two extensions into one, reading in reverse order; for which a better name would be the *collective rule*. For it to collect, the two extensions need a *c*ommon factor, which alphabetical order had fortuitously labeled *c*, better called the *collecting factor*.

There is a yet different way of labeling this rule that focuses more appropriately—and more transparently—on its relation with addition: extension *respects summary* because extension with a sum is a sum of extensions.

This kind of relation is pervasive within mathematics; so there is already even different jargon for it, often reserved for unary operations: they are said to be "*linear*", a term that derives from the straight or flat graphs of the most-elementary such operations.

The *linear* locution has scaling implications too. For the present example, scaling extension's arguments scales its result like so: $(2a) \cdot b = 2(a \cdot b) = a \cdot (2b)$. This rule is actually a consequence of respect for summary for smooth functions (like extension). For example, $(2a) \cdot b = (a+a) \cdot b = (a \cdot b) + (a \cdot b) = 2(a \cdot b)$; and similarly for $a \cdot (2b)$. This scaling rule propagates to arbitrary integers, rational numbers, and thence to real ones via limits.

Consequently, in this book the term *respect for summary* shall automatically encompass its induced *respect for scaling*. These evocative terms are not restricted to unary functions, as "*linear*" often is; and they are more accurate: graphs of non-elementary summary-respecting functions seldom have linear manifestations.

"Linear" differential equations, for example, have graphs

that are anything but straight. They would be better called *summary-respecting* equations so that smart students could immediately grasp their province.

A really smart student in third grade could probably grasp the meaning of *summary respecting* and *scaling respecting*; but—as every professor knows—it takes a really clever university math student to understand that "*linear*" refers not to straight graphs as it seems to, but rather to summary-respecting relations.

In sum, *summary-respecting* is usually the best way to describe extension's relation with addition because it is transparent and encompasses respect for scaling; but *collective* is more appropriate when collection is the narrow focus, as it is for adding bound vectors.

(¿Did you notice how I just went on and on? That is only because evocative terminology seems important to me.)

Adding bound vectors is all about exploiting their common collecting factor, which is just their intersection. Here is a picture for two different intersections, point c and free-vector c:

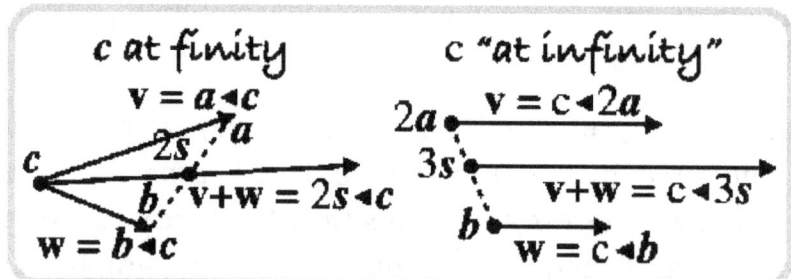

Adding bound vectors using collecting factors.

The intersection point on the left is point c at finity; on the right it is free vector c "at infinity". Let us walk thru the finite intersection first. There are two bound vectors, $\mathbf{v} = \mathbf{a} \triangleleft \mathbf{c}$ and $\mathbf{w}$

45

$= b \blacktriangleleft c$, whose collecting factor is point c. Whence …

$$\mathbf{v} + \mathbf{w} = a \blacktriangleleft c + b \blacktriangleleft c = (a+b) \blacktriangleleft c = 2s \blacktriangleleft c$$

… where $2s$ is the sum of $a+b$. Notice that the 2 in that *point* sum scales the *bound vector* by that amount: $2s \blacktriangleleft c = 2(s \blacktriangleleft c)$, owing to respect for summary.

It automatically induces the so-called "*parallelogram rule*" of vector addition, as you see. It would better be called the *tail-on-tail rule* because that ploy co-locates collecting factors, as the algebra requires. And so would head-on-head, if the collecting factors were at the heads. However, the head-on-tail ploy used for free vectors will not work for bound ones — it mis-locates their sum; give it a try.

(Incidentally, the tail-on-tail ploy, or the head-on-head one — meaning the parallelogram rule — also works point-wise for free vectors, but it does not generate a free vector in unitary form. Instead it generates a head–tail pair half as far apart with weights of 2, –2 (same formal separation). Do you see why? If so, you may suddenly realize that nothing-on-nothing also works the same way, meaning that free vectors may be added point-wise wherever they fell to earth. Conventional vectors cannot do that — they inherently add head on tail, as we shall see shortly.)

The bound-vector sum shown on the right intersects "at infinity" in free vector c, a collecting factor placed on the other side for variety (and for more intuitive extension). In this case $\mathbf{v} = c \blacktriangleleft 2a$ and $\mathbf{w} = c \blacktriangleleft b$ whence …

$$\mathbf{v} + \mathbf{w} = c \blacktriangleleft 2a + c \blacktriangleleft b = c \blacktriangleleft (2a+b) = 3(c \blacktriangleleft s)$$

… where $3s$ is the sum of $2a+b$. Again, notice that the 3 in that *point sum* scales the *bound-vector* by that amount.

However, this case fails to induce any conventional-vector rule because conventional parallel vectors are not fixed relative to each other.

The parallel ***well-located*** vectors (meaning ***bound*** ones) in the full geometric algebra are, and they intersect *algebraically* in a familiar "point at infinity" of projective geometry. That intersection "point" is nothing more than a free vector at finity. Its parallel rovability there implicitly manifests the dwindling of its sum to infinity.

Such an intersecting free vector would typically be normalized to have unit separation, like an intersecting point is typically normalized to have unit weight. However for projective purposes its separation is sometimes just ignored, as is the weight of a point.

Lines in physical space are seldom parallel, and they seldom intersect; so a sum of bound vectors there is usually irreducible, apparently having intrinsically composite dimension $\{2, 2\}$. However, that dimension is not in a unique minimal form; which, in physical space, is actually $\{2, 2\}$. This will become clear after we learn about elements with dimension $\{2\}$, namely …

Free bivectors

They provide our second hint that free elements in the full geometric algebra actually constitute a *sub-algebra* of it, meaning that they always generate a free result. Our first hint was the sum of two *free* vectors, u+v say, which produces another *free* vector. We are now going to see that an extension of those *free* vectors, u◄v, produces a *free* bivector.

(Bound elements do not constitute a sub-algebra because

they sometimes generate a free result, as with point subtraction. Moreover, altho a point extended with a point did generate a bound vector, so did a point extended with a free vector. Bound arguments are less tidy than free ones.)

Suppose we represent u as *a–b* and v as *c–d*. An easy way to extend them is to express u ◂ v as u ◂ (*c–d*), which generates u ◂ *c* – u ◂ *d*. Composed like that, we may detach u from *a–b* and place its tail on point *c*, and then on *d*, before extending. That gives two *separate-but-otherwise-exactly-opposite* bound vectors added together (remember, a dashed line indicates addition):

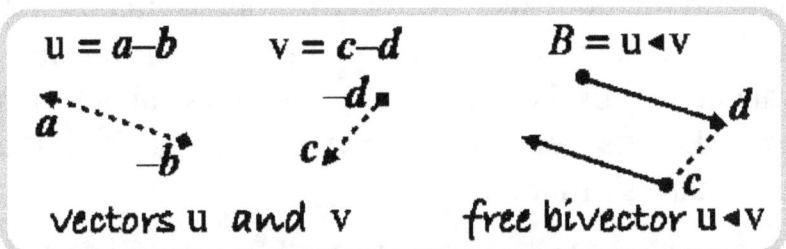

Free extension u ◂ v *by placing* u's *tail on c then d.*

A person might suppose that these bound vectors—since they are u-parallel—would coalesce to a single result like the previous c-parallel vectors did, which collected c into extension with a point. However, u here has been collected into extension with *c–d*, which fails to coalesce.

If *c–d* were approximated as done in the previous chapter, its extension with u at each step would generate a length-halved bound vector that would scoot twice as far away. In the limit that bound vector would vanish to infinity, whence one might suppose it becomes a classical "line at infinity". It actually becomes nothing at infinity, just as ill-defined and useless as the previous nothings there were.

Consequently, the sum of u ◂ *c* – u ◂ *d* does not exist within the algebra—this sum is irreducible, having intrinsically composite dimension {**2**, **2**}-*without-magnitude*, more conveniently abbreviated to free dimension {*2*}. Clearly, this bundle is free to move parallel to itself because each of the free vectors that generated it is. Strangely however, the dual *bondage* within that bundle gives it another kind of *freedom*, namely …

Shape-shifting ability

This ability *emerges naturally within a point algebra*, whereas it must be *artificially imposed upon a vector algebra*; in an interpretation, like vector freedom is imposed there. About that freedom, a point algebra had *explicit comment*; and also about vector bondage, and just now about bivector freedom. Let us see what further comment it has about bivector shape shifting.

The first thing it says is that the bound vectors within a free bivector may move anywhere within their confining lines; already a limited kind of shape shifting. The second thing it says is that we may compose u ◂ v differently; not top–bottom, but left–right as (*a–b*) ◂ v, which generates *a* ◂ v – *b* ◂ v. Here is a picture:

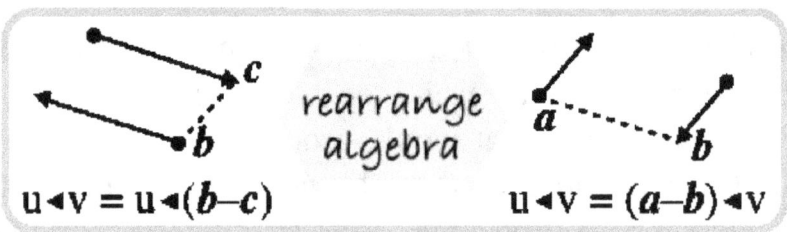

$$u \blacktriangleleft v = u \blacktriangleleft (b{-}c) \qquad\qquad u \blacktriangleleft v = (a{-}b) \blacktriangleleft v$$

A free bivector composed top–bottom and left–right.

These different compositions generate the *same* free bivector: they reside in the *same* plane; they have the *same* orientation there (clockwise viewed from this side); and they have the *same* area-separation — *same* free bivector.

You may skew it a little top–bottom, then recompose and skew a little left–right, then a little top–bottom, left–right … and so on. If you play with this you can make it tall and skinny, short and fat, rectangular, perfectly square, rhomboid; you can translate it anywhere you want, rotate it however you want …

But notice what you cannot do: you cannot change the bivector's *area-separation*, nor can you change its *planar-direction*, meaning the oriented plane it is parallel to. Those are invariants of any free bivector B; so it is often useful to explicitly decompose B into its scalar *separation* times its unit *direction*:

$$B = bI$$

Similarly, every other *free element* can be expressed as a separation times a direction. A *free element* is simply a free vector, or an extension of them; so it is intrinsically composite as a sum. For the *primitive* free elements, *separation*direction* is specifically *length-separation* times a *unit free vector*: $v = vi$.

(Notice the contrast in terminology between bound primitives and free ones: A bare "point" is implicitly a unit, tho it may have other primitive magnitude, namely weight; and specifically weight of −1, which makes it a "negative point". By contrast, a bare "vector" has no unit implications about its length; and the term "negative vector" is virtually never seen because vectors can be spatially reversed to remove a negative sign. So, even tho geometric formalities so far have failed to recognize the primacy of points, the

50

terminology of mathematics has long done so.)

Tho intrinsically composite, a free-element's lack of magnitude makes its *separate-but-exactly-opposite* bound summands cohere so well that they are better viewed as a kind of *coalescence* to a single roving thing. Happily, a free-as-possible basis does just that by bundling freedom right at the foundations, a *very* fertile idea we shall scrutinize shortly.

Adding free bivectors

Free bivectors have the same numeric dimension as bound vectors, namely two; so they can coalesce with them under addition, and also with themselves. This is analogous to free vectors, which have the same numeric dimension as points; so they can coalesce with them, and also with themselves.

Pursuing that analogy you might suppose that, under addition, a free bivector could translate a bound vector via end cancellation; and that it would coalesce with another free bivector, also via end cancellation. You would be mostly right, in physical space anyway, but free bivectors have shape-shifting nuances that free vectors lack. The simplest shaping arises in bound-vector translation:

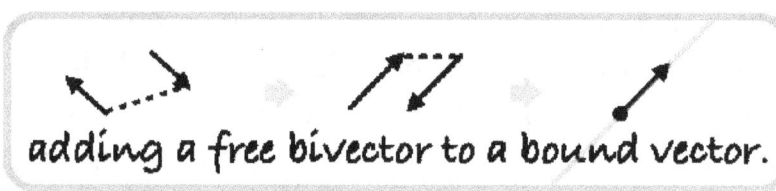

adding a free bivector to a bound vector.

Shape-shifting a free bivector for addition.

On the left you see a free bivector that is being added to the bound vector on the right, which lies within its plane. In its current shape and location, neither of its ends is able to annihilate the bound vector; so the first step is to shape it so one end is exactly opposite to that vector, indicated after the first light gray arrow. The next step is to translate the reshaped bivector to locate that end over that bound vector, indicated after the second arrow.

The final step is to add, which will annihilate the co-located vectors, leaving the left end of the reshaped-and-relocated bivector as residue. So, in this case the bound vector was effectively translated left-ward to itself by a distance equal to *bivector-area-separation / bound-vector-length*.

Clearly, this translation will only work if the bound vector lies within the plane of the free bivector. If not, then this sum is irreducible with dimension $\{2, 2\}$.

That dimension is in minimal form, but its summands will seldom be. In physical space, the magnitude*directed-locality of a bound vector, and the separation*direction of a free bivector can interchange under addition, owing to basis overlap. A particular balance between them will generate a free bivector perpendicular to the bound one, which is a unique minimal form. Happily, a free-as-possible basis makes this trivial, as we shall soon see.

Shape-shifted end-canceling also allows two free bivectors to coalesce to one bivector under addition. Again, this owes to basis overlap in physical space, which guarantees coalescence. This is obvious geometrically: the planes thru any two free bivectors intersect in a line at least, so two bivectors can always be shape-shifted to have their ends cancel on that line. Here is a picture:

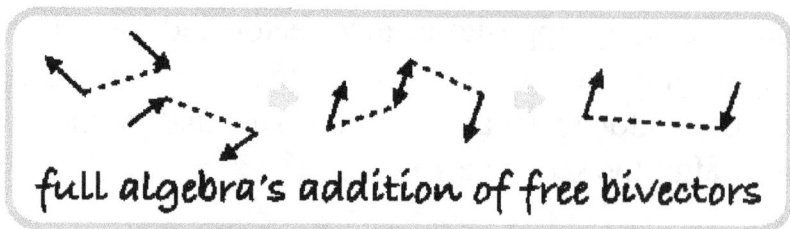

Shape-shifting free bivectors to add them.

On the left you see, in perspective, two free bivectors to be added. Each one is shape-shifted to have exactly-canceling ends on their line of intersection, as shown by the double-headed arrow in the middle figure. Those ends annihilate each other under addition, leaving two *separate-but-exactly-opposite* bound vectors added together—another free bivector —shown on the right. This is our third hint that free things constitute a sub-algebra: a *free* bivector plus a *free* bivector generates another *free* bivector via end-cancellation.

End-cancellation is the actual geometric process by which free bivectors add in the full geometric algebra; but I must emphasize again that the algebra shall automatically do the shape-shifted end-canceling for us when it is given a free-as-possible basis. $\hat{\ }_{+}\hat{\ }$

Such a basis, minus its anchor point, is the same one used in conventional geometric algebra. Unfortunately, that lack of an origin allows (free) vectors to be misinterpreted as line segments, and (free) bivectors as patches of plane. It shall be enlightening to observe the confusion this causes:

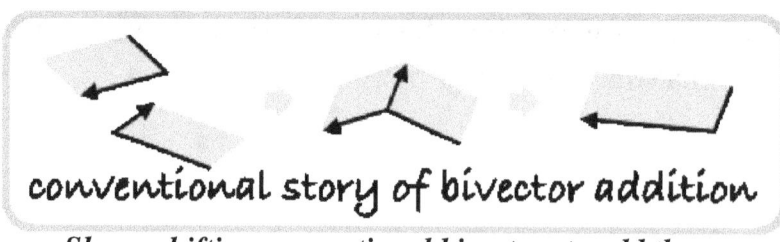

Shape-shifting conventional bivectors to add them.

The first step is similar: Each bivector is shape shifted to have a tail vector of one engage a head vector of the other. (This is rather cavalier in the conventional algebra, which does not articulate shape shifting like the full algebra does.) After the engagement, the conjoined areas magically *sprong* flat under addition, shown on the right.

It is the same story with vectors conjoined head-on-tail: they magically *sprong* straight after head engages tail. Unfortunately, there is no actual geometric process that *sprongs* conjoined lengths straight, or areas flat.

There *is* an actual geometric process that coalesces conjoined lengths to one length. We already saw it for two actual lengths, meaning bound vectors, which coalesce via a common collecting factor.

That shall work the same way for two actual areas, meaning bound bivectors, as we shall soon see. If they intersect in a line "at infinity", then they add via a free-bivector collecting factor. If they intersect in a line at finity, then they add via a bound-vector collecting factor. That induces a generalized parallelogram rule, rather than the generalized head-on-tail ploy just shown. The head-on-tail ploy would mis-locate this bound-bivector sum the same way it mis-locates a bound-vector sum.

Now that we know how to add free bivectors to themselves, and to bound vectors within their planes; and how to add bound vectors that intersect at finity or "at infinity", we are finally prepared to begin …

Adding skewed bound vectors

Such vectors do not intersect, not even "at infinity", so their sum is irreducible, apparently having intrinsically composite dimension {**2, 2**}. This dimension is not in minimal form, as mentioned, and we shall now see why.

The confining lines of two skewed bound vectors have a point on each that is closest to the other line. The geometric strategy is to slide each vector so that its tail is on that closest point; and then decompose them into a sum of parallel and anti-parallel vectors, as shown here:

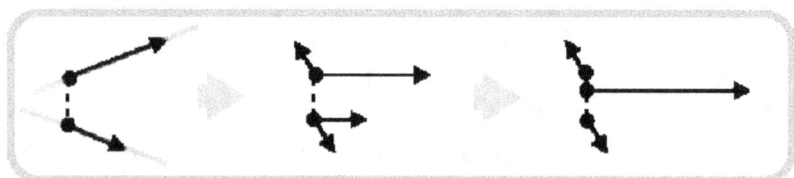

Making a skewed bound-vector sum minimal.

On the left you see two bound vectors to be added together, slid so their tails are as close as possible. In the middle, each has been decomposed into parallel and anti-parallel bound vectors. The anti-parallel ones are made perpendicular to the parallel vectors; and also made *exactly* anti, meaning they have the same length. On the right all of that has been added, giving a long bound-vector *thrust* pointing right, and a free-bivector *twist* perpendicular to it.

This sum is in minimal form and clearly has intrinsically composite dimension {**2, 2**}. Notice that the free bivector did not come out centered on the axis of thrust. It can, in fact, be placed anywhere owing to its freedom. The perhaps surprising implication for torquing operations is that a *twisting* torque can be offset from the fixed axis of thrust with no effect.

When decomposing the bound vectors, we need not have made the anti-parallel vectors perpendicular to the parallel ones; and we need not have made them exactly anti either. Of course we would not have produced a minimal form; but it is worthwhile exploring these alternatives to get a feel for the equivalence class of a sum of skewed bound vectors, which is huge.

The easiest way to explore is to notice that the dashed addition connecting closest points is perpendicular to each confining line. Therefor, viewing along that dashed line from above makes the bound vectors appear to intersect at their tails, an illusion a side view would dispel. Here is such a top view for various scenarios:

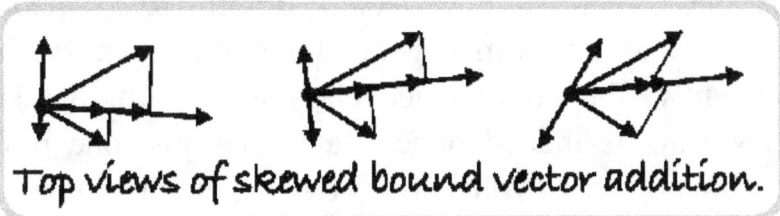

Top views of skewed bound vector addition.

Recomposing the sum of skewed bound vectors.

Let us orient ourselves with the middle figure—it is merely a top view of the minimal form already shown. You see the original vectors pointing up right and down right. You see their two parallel decompositions pointing right, which sum to the longer right-pointing bound-vector thrust. Finally, you see the two exactly anti-parallel bound vectors perpendicular to the thrust. Their separation generates the free bivector twist. Thrust and twist are the take-away from this figure.

The left figure relaxes the requirement for the anti-parallel vectors to be *exactly* anti. In consequence, they generate a long right-pointing bound vector plus a bound vector pointing up perpendicular to it, and positioned somewhat below it. Those two bound vectors are the take-away here, meaning that this ploy merely recomposes two skewed bound vectors into two other skewed bound vectors, whose sum is equivalent.

The right figure relaxes the requirement for the exactly anti-parallel vectors to be perpendicular to the parallel ones. The consequence is that they acquire more area separation; which strangely has no effect on the right-pointing bound-vector thrust. Altho their separation is larger, its projection perpendicular to the thrust remains the same—that is the part of the twist that has torquing significance along the thrust. That is the take-away here.

There are many other alternatives: we could have made the anti-parallel vectors not exactly anti *and also* not perpendicular either, which generates countless possibilities. Consequently, the equivalence class for a sum of skewed bound vectors is indeed huge. But it has just one minimal form: bound thrust plus free twist perpendicular to it, dimension $\{2, 2\}$.

¿Are you beginning to despair about the intricacies of recomposed minimal forms? Of shape-shifted canceling end-on-end free addition? Of tail-on-tail collecting bound addition? Of head-on-tail canceling free addition? How will you ever get the algebra to perform such intricate maneuvers? Happily, you don't need to—*the algebra will automatically do the intricate geometry* for you if you give it an anchored free-as-possible basis.

Free-as-possible joy

This will be brief, unlike my first several dozen efforts decades previous when I waxed lyrical about the expressiveness of a free-as-possible basis. True, it has much to wax about; but in my final pass I realized that my efforts had been half muck, half treasure; so here is the treasure, I think:

The *conceptual* advantages of a free-as-possible basis are that it directly manifests both freedom and shape in a transparent way, something a point basis utterly fails at. Consider, for example the *o p q* point basis for a plane.

A free vector expressed in this basis *indirectly* manifests its freedom by having basis weights that sum to zero. However, the basis lengths for a free bivector do not similarly manifest freedom, and on up. This may be seen by generating the free bivector $(p{-}o) \triangleleft (q{-}o)$, which equals $o \triangleleft p + p \triangleleft q + q \triangleleft o$.

Free vector p = $p{-}o$ has basis weights 1, −1, which sum to zero, easy enough to discern; and similarly for q = $q{-}o$. However, the free bivector they generate has basis lengths 1, 1, 1 that do not sum to zero. The fact that they generate something free requires calculation.

By contrast, that bivector expressed in a free-as-possible basis, p $\triangleleft$ q, requires no calculation at all to see that it is free— it obviously is because it is composed of free vectors.

Moreover, its shape is obvious too once you understand that it *effectively* engenders two separate-but-opposite bound vectors along its separations. The tri-sum $o \triangleleft p + p \triangleleft q + q \triangleleft o$ requires some enlightened sketching to arrive at that shape.

(Here is a place I am hoping you are not pencil-deprived. We will be walking thru this in detail in a dozen pages, but pre-effort is always preferable.)

The main *computational* advantages of a free-as-possible basis are that it *formally* distinguishes free from bound; and it never burdens the very versatile free sub-algebra with irrelevant bound computation, as a point basis would.

However, when bound computation is necessary, a free-as-possible basis rather unexpectedly represents each bound element in a simple uniform way as an ***origin-bound*** element plus a free translator.

To demonstrate, let us generate a bound vector **v** in the x y plane by extending its free part v from some translated point, o+p; where p is a classical "position vector", persuaded here to actually generate a position by translating the origin. This extension generates an origin-bound vector, $\mathbf{v}_o = \text{v} \triangleleft o$, plus a free bivector, $B = \text{v} \triangleleft \text{p}$. Here is a picture:

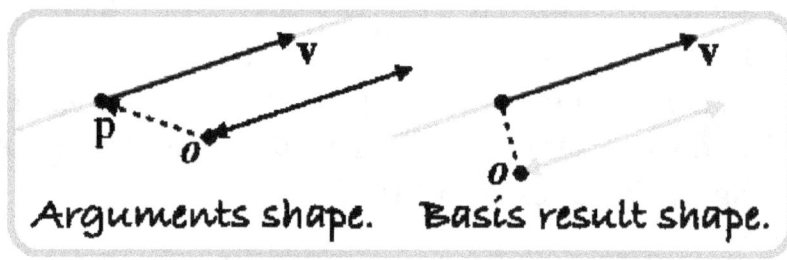

A free vector extended from a point: bound v.

Please slow down now—the following analysis is intricate-but-important. On the left you see, displayed in terms of the original arguments, free bivector B superimposed over origin-bound vector $\mathbf{v}_o$. The bottom end of the bivector exactly cancels that bound vector, as the oppositely-pointing heads indicate.

Those two oppositely-directed vectors annihilate each other under addition, leaving as residue the top end of the bivector, namely bound vector **v**, shown on top. This extension has two remarkable properties:

First, any position vector p will work that translates the origin to **v**'s confining line—such a vector always generates the *same* bivector v‹p, meaning that it always generates the same area-*separation* and the same planar *direction*, as should be clear. This is another way of demonstrating that bound **v** can move anywhere within the line thru itself because the shape of v‹p is irrelevant.

Second, shape irrelevance is automatically enforced by converting all possible v‹p bivector shapes into the same unique scaled version of y‹x, which is how free *B* manifests itself in the *o* x y basis. Consequently, this basis loses information about p's *absolute* position; but it does retain some perhaps surprising information about p's *relative* position—relative to free vector v, like so:

The basis retains information about free v via its scaling of x‹*o* and y‹*o*, the basis components of $\mathbf{v}_o$: free v is nothing more than the *free part* of $\mathbf{v}_o$, readily extracted simply by extricating the origin *o*.

Having extracted free-part v, we may retract it with bivector *B* (next chapter). That shall generate a vector perpendicular to v whose separation is simply the minimal distance from the origin to bound **v**'s confining line. This is the *rejection* of p from v, meaning what is left after the *projection* of p onto v has been removed. It implicitly establishes an automatic minimal form for all bound *n*-vectors.

For the case in mind, that minimal form is succinctly displayed on the right. On the bottom you see $\mathbf{v}_o$, pointing right. The bottom end of free bivector *B*, pointing left, has been superimposed over it, and grayed out since these bound vectors annihilate each other under addition.

That leaves the top end of the bivector as residue, bound

vector **v**, the ultimate result of v◂(*o*+p). Its minimal distance from the origin is the perpendicular distance indicated by the dashed line.

All bound *n*-vectors are represented in an anchored basis in the same way; as a free *n*-vector bound thru the origin, and then translated by a free (*n*+1)-vector. (So each basis element, bound and free, has numeric dimension *n*+1—this is an *n*+1-*composition*.)

Such uniform minimality makes addition of bound elements almost trivial in a free-as-possible basis. However, extricating the *elements*—the extensions of primitives—from that basis is sometimes far from trivial. The *Adding and Composing* chapter shall rassle with it.

For now, let us proceed with trivial elements by adding the two intersecting bound vectors previously done in a point basis. Here is the corresponding free-as-possible figure:

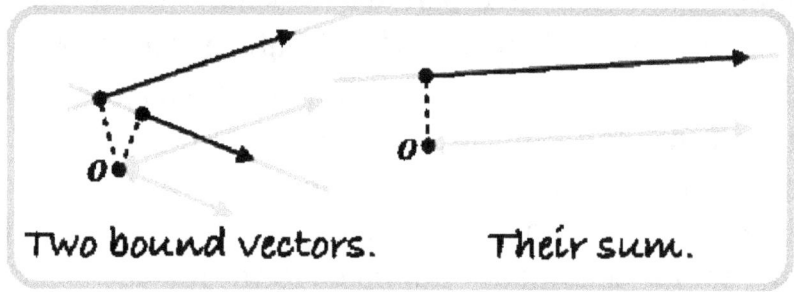

Two bound vectors.   Their sum.

Free-as-possible basis representation.

On the left you see the previous two bound vectors, but now represented in a free-as-possible basis. (The top vector accidentally has its tail at the intersection of the confining lines owing to an unexpected coincidence in the previous point-based figure. I just left it there—I didn't have time or patience to go back and revise that intricate figure. Ha—I am emulating my hero Grassmann.)

In minimal form, tail points shall be displayed at minimal distance from the origin, and this will seldom be an intersection point, if there is one. A bound element is always first bound thru the origin by its bound basis elements, as just seen. It then gets translated from there via end-canceling of its free basis elements, as the double-headed light gray arrows on the bottom indicate.

Notice that, in order for these bound vectors to intersect after translation, each of their translation bivectors must be parallel to the plane thru *both* bound vectors at the origin. Otherwise the vectors will diverge during translation, so they will not intersect after it.

Finding such intersection was necessary in the point algebra in order to engage the collective rule. That rule is already implicit in a free-as-possible basis owing to common initial intersection thru the origin—the vectors are simply collected together by addition of their scaled origin-bound bases.

Then the free bivectors are collected in the same way; which implicitly adds via end-canceling, as just explained. The result is a bound vector translated from the origin to its confining line, as you see on the right.

Enlightening exercise: Perform the same graphic analysis in minimal form on the parallel bound vectors previously added via their intersection "at infinity". Their initial parallel bondage thru the origin will give them all the same confining line. However, their subsequent translations do not require parallel free bivector translators, as the previous at-finity case did. They merely require those translators to intersect on the common confining line thru the origin.

Finally, let us make our free-as-possible analysis less trivial: let us reconsider the previous bound-as-possible analysis of skewed bound vectors, which do not intersect anywhere: not at finity; not even "at infinity". In such a case they will start bound thru the origin, as usual, but non-parallel. Their free bivector translators will also be non-parallel—this is necessary so that the bound vectors will diverge during translation, and will not intersect after it.

In consequence, their addition will generate a vector sum bound thru the origin; but a free bivector sum that is *not parallel* to that bound vector result. Such a bivector cannot translate it by end canceling, as done in the previous two examples.

Fortunately that free bivector can be factored into a part that *is* parallel to the origin-bound vector, and a part perpendicular. The parallel part can then translate the origin-bound vector to its confining line, as before. This gives the bound *thrust*, previously laboriously derived point-wise. The residual perpendicular part of the bivector is simply the free *twist*.

A canonical bound thrust plus a free twist, dimension {**2**, *2*}, is the most general two-dimensioned composition in physical space—bound **4**-space. Addition of such 2-compositions always generates another such; but of course the twist may vanish, leaving bound dimension {**2**}; or the thrust may vanish, leaving free dimension {*2*}; or both may vanish, leaving dimension {}, meaning 0.

In higher spaces, 2-compositions may become more intricate. In the next higher space for example, bound **5**-space, with basis primitives *o* w x y z, free bivectors w◄x and y◄z do not have any basis overlap, so their sum is intrinsically composite, having minimal dimension {*2, 2*}, a dimension that is not valid in physical space.

This preview of a free-as-possible anchored basis was intended to help you start appreciating how the algebra actually works; but such a basis will not really be understood until you absorb the ideas in the *Retracting* chapter. You will then be prepared to extricate elements in the *Adding and Composing* chapter; after which you will be ready for *Synthesis* of extension with retraction. Let us advance toward those goals by first mastering 3-compositions.

Bound bivectors

Here is where we finally arrive at bona-fide areas, dimension {**3**}, generated by extending three independent points, $a \cdot b \cdot c$ say. Immediately there is a novelty: Asked algebraically, ¿How do we group these extensions? $(a \cdot b) \cdot c$ or $a \cdot (b \cdot c)$? Asked geometrically, ¿Which extension do we do first? $a \cdot b$ or $b \cdot c$? The answer is that it does not matter, as you see:

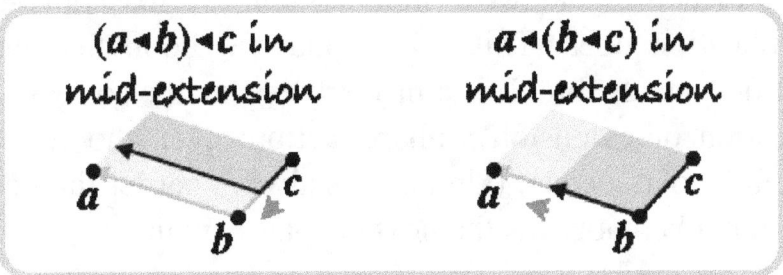

Validating extension's associative law geometrically.

On the left you see bound vector $a \cdot b$ being *extended from* c back to its original position. On the right you see bound vector $b \cdot c$ being *extended to* a from its original position, reading that extension backward.

Grouped extensions can be read either forward or backward —the sweeping process is the same. To verify this, perform the grouped sweeping in the last extension in the conventional way as $(c \wedge b) \wedge a$ (reading forward of course—the bilateral wedge $\wedge$ fails to suggest that extensions might even be read backward). That is why we are able, for vector consistency, to uniformly *extend from*.

Notice that in an sequence of point extensions, the very first extension *from* a point generates a *tail* there. The next extension sweeps that by its *tail from* the next point, thereby generating a new tail. And so on. Conversely, the very first extension *to* a point generates a *head* there. The next extension sweeps that by its *head to* the next point, thereby generating a new head.

In a nutshell: *from* extension sweeps *tail-from* a point back to its original position; *to* extension sweeps *head-to* a point forward from its original position. Peering carefully at the previous figure should clarify this.

Each differently grouped extension produces the same result, meaning that extension *associates*, like third-grade multiplication does. This is a wonderful boon that greatly simplifies the algebra. It is a property we shall also seek in the unification of extension with retraction; even tho retraction itself fails to associate. (Undoing is always harder than doing, even in mathematics, as the next chapter explains.)

There is another novelty that bound bivectors present: extension's neg-commuting is valid even for extension factors that are not adjacent to each other. Suppose, for example, that we want to exchange the first and last factors in $a \cdot b \cdot c$.

We can walk a over to where c had been: $a \cdot b \cdot c = -b \cdot a \cdot c$

65

= $b \cdot c \cdot a$. Then we can walk c back to where a had been. Coming back requires one less swap than going did because the last going-swap already moved c back one place: $b \cdot c \cdot a$ = $-c \cdot b \cdot a$; we have finally exchanged a with c, a negation.

It works the same for extension factors separated by *any n* places: walking the first factor to where the second had been requires n swaps. The last swap moves the second factor one place closer to the first; so coming back requires $n-1$ swaps. That is $2n-1$, always an odd number of swaps, an odd number of negations, equivalent to a single negation. So two factors anywhere within an extension neg-commute.

Speaking of neg-commuting, it is actually relatively rare among entire elements themselves. For example, a bound vector commutes with a point under extension: simply walk the point across the bound vector's two extension points. That is two negations; effectively none at all. In fact a bound vector commutes with *anything* under extension: walk each primitive factor across the bound vector, one after the other. Each trip leaves sign unchanged, so the entire journey does too.

Illuminating exercise: The previous paragraph developed the *algebra* of bound-vector commuting, but you may be surprised to see its *geometry*. So try a sketch of bound vector $a \cdot b$ extended ***to*** and ***from*** point c. Do these bound bivectors have the same shape? If not, are they really the same? Can they be given the same shape by sliding the bound vector along its confining line before extension?

Not just bound vectors, but *any* even-dimensioned element, bound or free, commutes with *anything* under extension for the same reason: Walk each factor across it. Each trip leaves sign unchanged.

That leaves only an odd-dimensioned element extended with another one. Here we finally achieve neg-commuting: walking a factor across that element negates on each trip; and there are an odd number of trips, an odd number of negations, equivalent to a single negation.

This commuting analysis is simple when made in terms of numeric dimension, but would be almost impossibly complex if attempted on the n-value of an n-vector—free and bound ones have opposite dimensional parity.

Just remember, n-value specifies *spatial expanse*, not extension dimension. This idea initially confused me and most other new-comers to geometric algebra.

Incidentally, since scalars have even dimension {0}, they also commute with everything under extension; but what does that mean? Well, extension with a scalar cannot change other dimensions, so the only action left is to enlarge or diminish —*scale* in short.

In other words, extension with a scalar seems to be nothing more than scalar multiplication. Ponder its implications for a weighted point: $aa = a \triangleleft a = a \triangleleft a = aa$, which is to say …

*The **weight** of a point is simply its **free part**. Said conversely, a **weighted point** is nothing more than a **bound scalar**.*

This is a potent idea, but it needs to be disciplined by geometric meaning to avoid contradictions. Otherwise, we might be seduced into thinking we can unify extension and scalar multiplication into one operation.

For example, we might think we could write extension $aa \triangleleft bb$ as $a \triangleleft a \triangleleft b \triangleleft b$; which could be commuted and grouped like so: $(a \triangleleft b) \triangleleft (a \triangleleft b)$, which becomes $ab(a \triangleleft b)$. ¿So why even bother with the extension symbol? Since extension has become scalar multiplication, why not just use simple symbol juxtaposition $aabb = abab$ for both products?

That would indeed be the elegant syntactic way to proceed if extension and scalar multiplication were the only products in the algebra. But it would preclude undoing extension, namely retracting; and would also preclude their unification, extension-retraction. Worse, it would just be wrong, semantically.

The semantics of extension are that it is a *dimension-augmenting, located operation*. So the unlocated dimension-ignoring product $a \triangleleft b$ is not valid, and would in fact generate contradiction when deployed with retraction, as you shall see in the next chapter.

This has implications for *who is extending whom*. It usually does not matter: the big guy can extend the little guy or vice versa. With scalar extension, however, it is always the big *located* guy doing the extending. A little *unlocated* guy doing it has no meaning, and neither do two little guys doing it.

Here is the geometric distinction between these scalings: $2a$ is a special kind of addition, $a+a$; $2 \triangleleft a$ is a special kind of dimensional *ascent*—from dimension $\{0\}$ to dimension $\{1\}$. Fortuitously, these different forms happen to produce the same result, stand-alone.

However, when deployed in an extension-undoing equation, the dimensional-ascent form is crucial for dimensional *descent*, the only proper use of it. Retraction shall use it to descend weighted point aa to its free part a, among other uses. You'll see.

Shape shifting

Extension's associative rule allows us to slide the bound vectors within $a \triangleleft b \triangleleft c$ individually. We may slide $a \triangleleft b$ a little along its confining line; or slide $b \triangleleft c$ a little along *its* confining line. Here is a picture

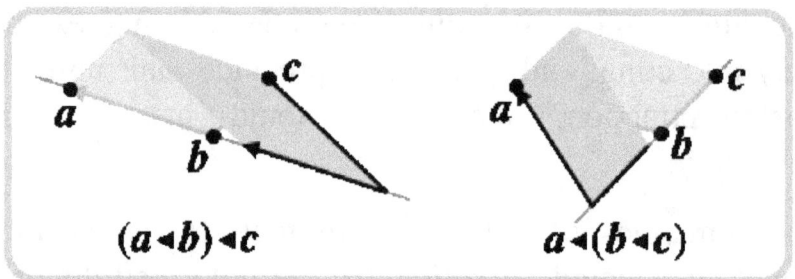

$(a \triangleleft b) \triangleleft c$ $a \triangleleft (b \triangleleft c)$

Skewing a bound bivector along confining lines.

However there is a problem: the corner point diagonal to b in $a \triangleleft b \triangleleft c$ is pinned down, as you see. Let us call that point d. It clearly lies within $a \triangleleft b \triangleleft c$'s plane, so it must be expressible in terms of those three points.

Indeed, the midpoint of b and d equals the midpoint of a and c, so $b+d = 2m = a+c$, whence $d = a+c-b$. (Enlightenment: compose this alternatively as a plus free vector $c-b$, or else as midpoint $2m$ minus b).

Having legitimized d within $a \triangleleft b \triangleleft c$'s plane, we may express that bound bivector in two other ways: $b \triangleleft c \triangleleft d$ or $c \triangleleft d \triangleleft a$, as you may check. That gives us two more confining lines to slide along: the one on the top thru $c \triangleleft d$ and the one on the left thru $d \triangleleft a$. (Neg-commuting actually provides two others, the diagonal one thru $b \triangleleft d$ and the diagonal one thru $a \triangleleft c$—do you see why? However, they provide no further shape-shifting ability.)

So, just as with a *free* bivector, we may skew this ***bound*** one a little top–bottom, then recompose and skew a little left–right, then a little top–bottom, left–right … and so on. Again, if you play with this you can make it tall and skinny, short and fat, rectangular, perfectly square, rhomboid; you can translate it anywhere you want within its plane, rotate it however you want there …

But notice what you cannot do: you cannot change its *area*, and you apparently cannot change its *directed-locality* either, meaning the oriented plane it lies within.

However, we need to verify that because free-bivector skewing seemed unable to change directed-locality either. Nonetheless, we knew it could change locality, at least, because the two free vectors that compose it could. The three points that compose $a \triangleleft b \triangleleft c$ seem to lack that ability; but is that really true?—composed points, we have seen, are often adventurous.

Bondage

Demonstrating bivector bondage is the same as demonstrating vector bondage—show that $a \triangleleft b \triangleleft c$ arises by extension of its free part with a point anywhere inside its confining plane, but never arises with a point outside it.

The free part of a bound bivector—as with the free part of a bound vector, or even the free part of a weighted point—is simply the free element that can generate it under extension with a unit point; a free bivector in this case.

And as with a bound vector, this is most easily done by placing one end of the free bivector over the point and then extending. Poof!—that end vanishes under extension, leaving a bivector bound thru that point. Here is a picture for free

70

bivector B.

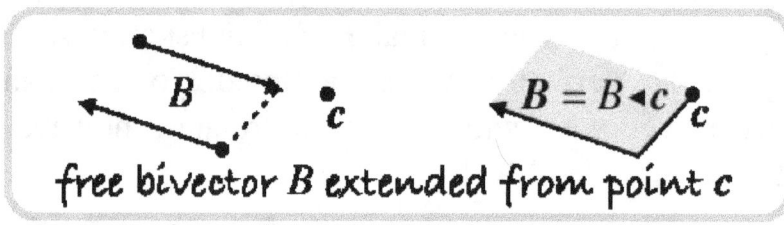

free bivector B extended from point c

Poof-extending a free bivector from a point.

On the left you see free bivector B next to point c. Place one end of it on that point before extending. That end will vanish because it generates no area, leaving the other end extended from that point, shown on the right. Our first job is to algebraically compose a free bivector B that is indeed the free part of bound B.

This is only slightly harder than composing the free part of bound v had been: generate an extension of two free vectors in terms of bound B's extension points, head on tail, in the same order like so: $(a–b) \cdot (b–c)$, which generates $a \cdot b + b \cdot c + c \cdot a$.

Very enlightening exercise: Make a sketch showing that the three bound vectors just generated induce a head-on-tail cycle around $a \cdot b \cdot c$'s extension points. Next make sketches showing—*in three different ways*—that this triple sum is indeed a free bivector, meaning that it reduces to a sum of two *separate-but-exactly-opposite* bound vectors. To do so, you will need to juxtapose tail points by sliding each bound vector, in succession, along its confining line.

Armed with new free-part B and your three sketches of it, we can begin extending it from some points. Let us start with point a: $B \cdot a = (a \cdot b + c \cdot a + b \cdot c) \cdot a = a \cdot b \cdot c = B$.

¿Did you see the magic?—extension with a ignores all

71

extensions containing that point because the repeated factor makes them vanish. That leaves $b \triangleleft c \triangleleft a$, which double neg-commuting converts into $a \triangleleft b \triangleleft c$. You really have to do this yourself, at least twice, to see the magic, for which here is an …

Enlightening exercise: Extend free B from points b and c. This is important. You will get bound B in both cases. This means that bound B is able to move anywhere within the plane thru itself. To see that, extend free B from any point $d = aa + bb + cc$ within that plane:

$$B \triangleleft d = B \triangleleft (aa+bb+cc) = aB+bB+cB = (a+b+c)B = 1B$$

Conversely, if d were not expressible in terms of a, b, c—if d resides outside that plane—then $B \triangleleft d$ would not equal $a \triangleleft b \triangleleft c$: that bivector is indeed bound to the plane thru itself.

Adding bound bivectors

This works the same as adding bound vectors: use the collective rule to coalesce two bound bivectors into one. In this case the collecting factor is their intersection in a line, either at finity in bound vector c; or "at infinity" in free bivector C. Here is a picture:

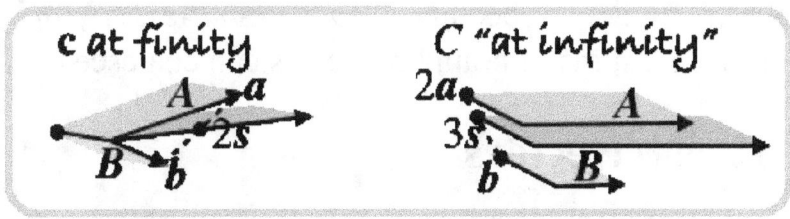

Adding bound bivectors intersecting in a line.

The bound-bivector sum shown on the left intersects at finity in bound vector c like so: $A = a \triangleleft c$ and $B = b \triangleleft c$, whence …

$$A+B = a \triangleleft c + b \triangleleft c = (a+b) \triangleleft c = 2(s \triangleleft c)$$

… where $2s$ is the sum of $a+b$. Again, the 2 in that *point* sum scales the *bound-bivector* by that amount, which induces a generalized parallelogram rule, as you see.

The bound-bivector sum shown on the right intersects "at infinity" in free bivector C, a collecting factor placed on the other side for variety: $A = C \triangleleft 2a$ and $B = C \triangleleft b$ whence …

$$A+B = C \triangleleft 2a + C \triangleleft b = C \triangleleft (2a+b) = 3(C \triangleleft s)$$

… where $3s$ is the sum of $2a+b$. And again, the 3 in that *point sum* scales the *bound bivector* by that amount, which induces the same kind of simple magnitude-adding already seen for parallel bound vectors.

For projective purposes, the intersection lines—c at finity, and C "at infinity"—might be normalized to units; but more likely the length of c and the separation of C would just be ignored.

We have now seen almost the entire scenario for adding bound bivectors in physical space. That space does not have enough elbow room for bound bivectors not to intersect, unlike the case for bound vectors, so their sum nearly always coalesces to another bound bivector, dimension {3}. However, under special *separate-but-exactly-opposite* circumstances, sums of bound bivectors can coalesce to …

Free trivectors

They provide our fourth *strong* suggestion that free elements in the full geometric algebra constitute a sub-algebra: extension of three *free* vectors, u ◄ v ◄ w, say, produces a *free* trivector, T, having *free* dimension {*3*}.

That dimension is an abbreviation for dimension {**3**, **3**}-*without-magnitude*, which tells us that a free trivector is a sum of *separate-but-exactly-opposite* bound bivectors. To see this, suppose that we factor T = u ◂ v ◂ w into a free bivector *B* = u ◂ v extended from a free vector, w = *a–b*. Composed like that, we may place one end of *B* on point *a*, and then on *b*, before extending. That gives two *separate-but-otherwise-exactly-opposite* bound bivectors added together as you see:

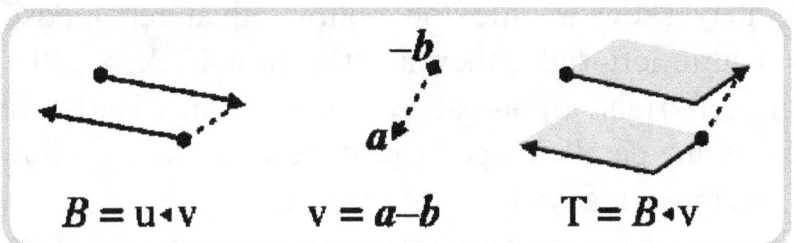

$$B = u \blacktriangleleft v \qquad v = a\!-\!b \qquad T = B \blacktriangleleft v$$

Free trivector: free bivector extended from a free vector.

You likely realize by now that this addition does not coalesce to a valid sum: If v = *a–b* were approximated as done previously, its extension with *B* at each step would generate an area-halved bound bivector that would scoot twice as far away.

In the limit that bound bivector sum would become nothing at infinity, just as ill-defined and useless as all the previous nothings there were. Consequently, free trivector T is indeed an irreducible roving bundle of summands having free dimension {*3*}.

You also likely realize that free T is just as much a shape-shifter as the previous free elements were—*more* of a shape-shifter because when the volume shape of its separation changes, the area shape of its faces also changes. Shape shifting arises just as it did with free bivectors: we may compose u ◂ v ◂ w pointwise not only on w, but also on u and v.

Here is a picture:

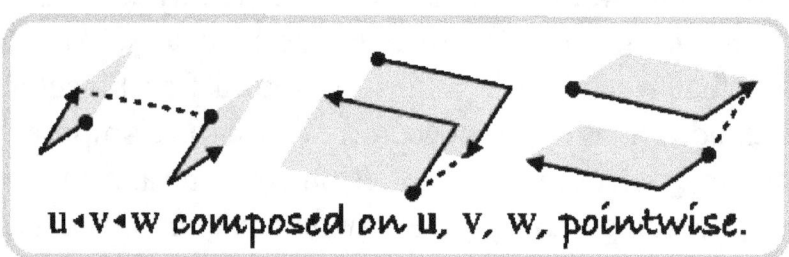

u ◄ v ◄ w composed on u, v, w, pointwise.

Free trivector variously composed pointwise.

We may skew a little top–bottom, then recompose and skew a little left–right, then a little front–back, a little top–bottom, left–right, front–back … and so on. Clearly we can make this any parallelepiped shape we want, rotate it however we want, move it anywhere we want …

But again, none of this skewing can change the trivector's *volume-separation*, nor can it change trivector *direction*, meaning the directed volume it is parallel to.

Understanding that requires more imagination than previous shape-shifting did. That volume — in both possible physical directions, right-handed and left-handed — is the only space available to us humans; meaning that our trivectors are embedded in it, so are automatically parallel to it.

Adding free trivectors

Free trivectors have the same numeric dimension as bound bivectors, namely three; so they can coalesce with them under addition, and also with themselves.

Having seen this scenario twice before, you might suppose that, under addition, a free trivector could translate a bound bivector via end cancellation; and that it would coalesce with another free trivector, also via end cancellation. You would be completely right, in physical space anyway; but of course there are shape-shifting nuances. The simplest arise in bound-bivector translation:

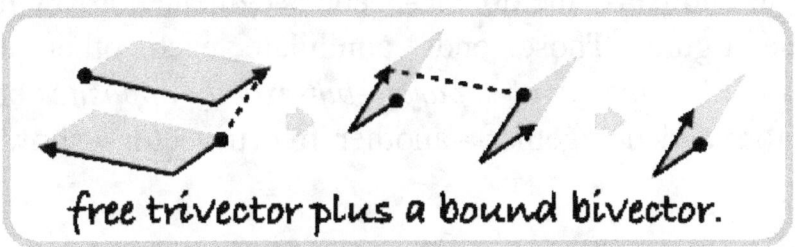

free trivector plus a bound bivector.

Shape-shifting a free trivector for addition.

On the left you see a free trivector that is being added to the bound bivector on the right, which lies within its volume (as it must in physical space). In its current shape and location, neither of its ends is able to annihilate the bound bivector; so the first step is to shape it so one end is exactly opposite to that bivector, indicated after the first light gray arrow. The next step is to translate that reshaped trivector to locate that end over that bound bivector, indicated after the second arrow.

The final step is to add, which will annihilate those co-located bivectors, leaving the left end of the reshaped-and-relocated trivector as residue. So the bound bivector was effectively translated left-ward to itself in this case by a distance equal to: *trivector-volume-separation / bound-bivector-area*. Which prepares us for end-canceling addition of free trivectors themselves:

addition of free trivectors in our space.

Shape-shifing free trivectors to add them.

On the left you see, in perspective, two free triivectors to be added. Each one is shape-shifted to have exactly-canceling ends, as shown by the double-headed segmented arrow in the middle figure. Those ends annihilate each other under addition, leaving two *separate-but-exactly-opposite* bound bivectors added together—another free trivector—shown on the right.

This is our fifth *very strong* suggestion that free things constitute a sub-algebra: a *free* trivector plus a *free* trivector generates another *free* trivector via shape-shifted end-canceling.

That maneuver is always the *geometric meaning* of free addition; but we have seen that a free-as-possible basis does it automatically for 1- and 2-compositions. Let us see how that works for 3-compositions.

More free-as-possible joy

An anchored free-as-possible primitive basis for physical space shall systematically induce the following extended three-dimensioned basis: $y \triangleleft x \triangleleft o$, $z \triangleleft x \triangleleft o$, $z \triangleleft y \triangleleft o$, $z \triangleleft y \triangleleft x$. Notice how unexpectedly **bound** this new *free*-as-possible 3-basis is—almost as bound-as-possible: three bound bivectors and one free trivector.

For that reason *only one* of these elements can be used to represent free trivectors, namely $z \triangleleft y \triangleleft x$—*any* free trivector is nothing more than a scaled version of this unit basis element. When enhanced by Clifford, this makes free trivectors behave like scalars, which are, after all, nothing more than scaled versions of unit 1. So free trivectors are often called *pseudo-*

scalars for physical space, focusing on the algebra; but I prefer *free ceiling*, focusing on the geometry.

Now let us switch to the other 3-compositions, bound bivectors. This shall be deja vu all over again. The most elegant way to generate a bound bivector B is to extend its free part B from some translated point, $o+p$, just as before. This extension generates origin-bound bivector $B_o = B \triangleleft o$ plus free trivector $T = B \triangleleft p$. Here is a picture:

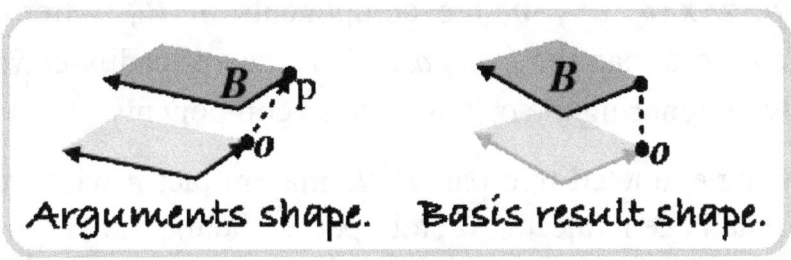

A free bivector extended from a point: bound B.

On the left you see—displayed in terms of the original arguments—free trivector T superimposed over origin-bound bivector B_o . The bottom end of the trivector exactly cancels that bound bivector, as the oppositely-pointing heads indicate. Those two oppositely-directed bound bivectors annihilate each other under addition, leaving as residue the top end of the trivector, namely bound bivector B. Again, this extension has two remarkable properties:

First, any position vector p will work that translates the origin to B's confining plane—such a vector always generates the *same* trivector $B \triangleleft p$, meaning that it always gives it the same volume *separation* and the same *direction*, as should be clear. (There are only two directions available for a volume in physical space, right-handed and left-handed.) This is another way of demonstrating that bound B can move anywhere within the plane thru itself because the shape of $B \triangleleft p$ is irrelevant.

Second, shape irrelevance is automatically enforced by converting all possible $B \cdot p$ trivector shapes into one scaled version of $z \cdot y \cdot x$, which is how free T manifests itself in the basis. Consequently, this basis loses information about p's *absolute* position; but it does retain information about p's *relative* position—relative to free bivector B, like so:

The basis retains information about free B via its scaling of $y \cdot x \cdot o$, $z \cdot x \cdot o$, $z \cdot y \cdot o$, the components of $\boldsymbol{B_o}$ —free B is nothing more than the *free part* of $\boldsymbol{B_o}$, again readily extracted simply by ignoring the origin in those components.

Having extracted free-part B we may retract it with free T. That shall generate a vector perpendicular to B whose separation is simply the minimal distance from the origin to bound $\boldsymbol{B}$'s confining plane. This is the *rejection* of p from B, which automatically establishes a minimal form for bound bivector $\boldsymbol{B}$.

That minimal form is succinctly displayed on the right. On the bottom you see $\boldsymbol{B_o}$, clockwise. The bottom end of free trivector T, counter-clockwise, has been superimposed over it, and grayed out since these bound bivectors annihilate each other under addition.

That leaves the top end of the trivector as residue, bound bivector $\boldsymbol{B}$, the ultimate result of $B \cdot (o+p)$. Its minimal distance from the origin is the perpendicular distance indicated by the dashed line.

Now that we know how a free-as-possible basis represents bound bivectors in minimal form, let us add two of them in that form, like we did two bound vectors. Here is a picture:

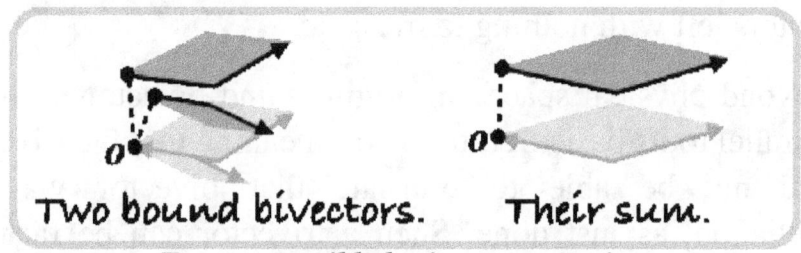

Two bound bivectors.   Their sum.

Free-as-possible basis representation.

On the bottom left you see two bivectors bound thru the origin. They have been grayed out because each has been canceled under addition by the bottom end of its free-trivector translator. That leaves the top end of each free trivector as residue, the two bound bivectors displayed in dark gray.

Each translator is shown squared-up in minimal form because the basis retains no information about original shape. Consequently each dashed-line addition is displayed perpendicular to its ends, causing each bound-bivector result to be positioned at its minimal distance to the origin. As mentioned, this distance may be explicitly extracted by retraction.

When these two bound bivectors are added, their origin-bound basis representations are added along with their free-trivector translators. The result is again displayed squared-up in minimal form, as you see on the right: the canceled origin-bound bottom is grayed out, the top residue is the ultimate bound-bivector result, in dark gray.

This case is atypical because, in physical space, the origin-bound bivectors are automatically parallel to their trivector translators, unlike the previous bound-vector case. Consequently, addition of two bound bivectors almost always generates another, dimension {**3**}. Only if their origin-bound bivectors are *exactly opposite* do they generate a different dimension, namely {*3*}, which is the sum of two naked free trivectors left with nothing to translate.

Beyond physical space, an origin-bound bivector need not be parallel to its basis free-trivector, meaning that the trivector would not be able to translate that bivector via end cancellation, as just done. Such a trivector can be factored into a part that *is* parallel to the origin-bound bivector, which does translate it, and a part perpendicular, just as done with bound vectors.

This would generate dimension {**3**, *3*} in the simplest case, but in higher non-intertwined spaces, 3-compositions could generate more intricate dimensions, as previously glimpsed for 2-compositions. This will be pondered in the *Adding and Composing* chapter. For now we shall ponder the last possible elements in physical space.

Bound trivectors

There is really only one bound trivector in physical space, up to scaling, because each one is represented in the basis as a scaled version of unit bound trivector $\mathbf{I} = z \triangleleft y \triangleleft x \triangleleft o$. This is simply *the* unit free trivector $I = z \triangleleft y \triangleleft x$ extended from the origin. Here is how to visualize that extension geometrically:

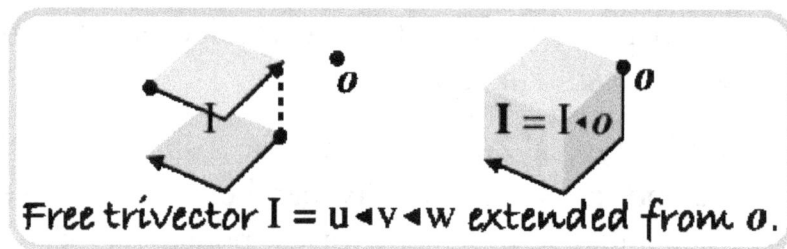

Free trivector $\mathbf{I} = u \triangleleft v \triangleleft w$ extended from o.

Free trivector extended from a point: bound trivector.

As usual, this free trivector has one end placed over the point before extension. Poof!—that end vanishes during extension because it generates no volume. This leaves the other end extended from the point, a bound trivector. Its shape-shifting bondage allows it to slide anywhere in physical space, so it is just as free there as a free trivector, or any other free element.

Notice that it can be more primitively and more perspicuously expressed in terms of points like so: **I** = $z◄y◄x◄o$, where x = *x–o*, y = *y–o*, z = *z–o*. You can easily check this by substituting these expressions into z◄y◄x◄o. This ploy is important—give it a try.

It is important because it demonstrates that any extension of free vectors (hence any free element) with at least one point effectively becomes a *pure point extension*, meaning a bound element—that point can be used as a tail point in each of the free vectors, which will eliminate all tail points, leaving only head points. Said more geometrically, extension of anything—free or bound—with a point binds it thru that point; as we have seen graphically over and over.

Now that bound trivector **I** has been expressed in terms of points, its free part is easy to construct algebraically: compose an extension of three free vectors in terms of bound **I**'s extension points, head on tail, in the same order, like so: $(z–y)◄(y–x)◄(x–o)$. This extension is a little tricky because of repeated extension factors, so it is an ...

Excellent exercise: Show that it generates $z◄y◄x - z◄y◄o + z◄x◄o - y◄x◄o$. Then make sketches showing that any pair of these four terms has a common collecting factor (a bound vector—a line of intersection) extended with a free vector. Whence the remaining pair does too. There are three ways to do this. Each way generates a pair of *separate-but-exactly-opposite* bound bivectors—a free trivector, I let's call it.

Next, extend free $I = z \cdot y \cdot x - z \cdot y \cdot o + z \cdot x \cdot o - y \cdot x \cdot o$ from each of its extension points in turn. You will almost magically get unit bound trivector **I** every time. Use that fact to show that extension from any unit point $c = xx + yy + zz + oo$ within the space thru **I** also generates it. Finally show that any point outside that space fails to generate **I**. This is moot in physical space, which does not have any points outside; but it is not moot in higher spaces, which do.

We shall not meander to higher spaces because physical space has already elucidated all the rules of point addition and extension. Let us pause to savor the free n-vectors and bound n-vectors these rules have produced so far. They are shown here in sky-blue clarity because we are not in Kansas any more—Grassmann's spatial arithmetic has finally moved us well beyond Euclid and Descartes:

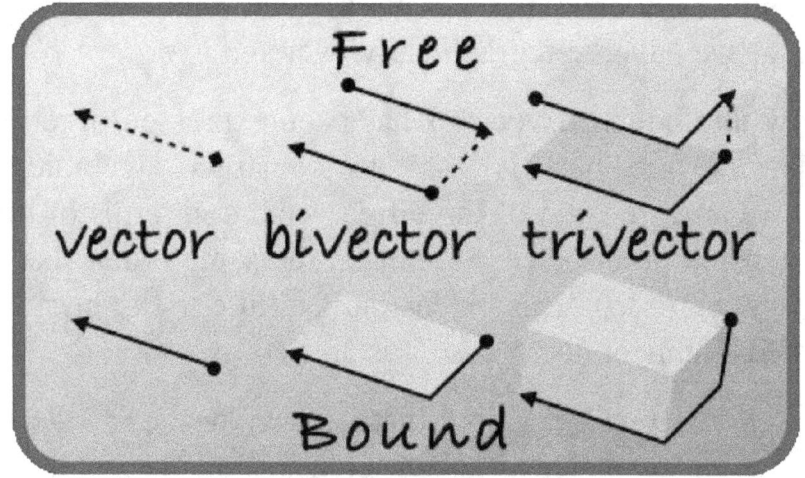

Expansive numbers in physical space.

Orientation

Travel advice: Curiously, your author is fascinated about orientation as enhanced by the free-versus-bound distinction, but you may not be. What follows may be more than you ever wanted to know.

Each of the *expansive numbers* displayed here has a physical orientation that is often expressed ambiguously by failing to specify what it is relative to. Bivector orientation is the classic example—¿What are the orientations of these bound bivectors?

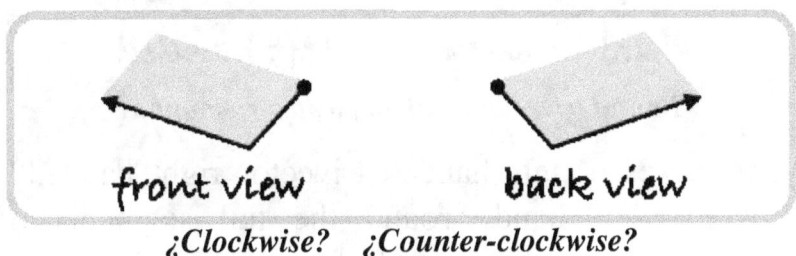

¿*Clockwise?*   ¿*Counter-clockwise?*

¿Is the left bivector clockwise? Is the right one counter-clockwise? In fact, they are the *same* bivector, viewed from the front and back. ¿But which is front, and which is back? It depends on your viewpoint. Until that is specified, bivectors have ambiguous orientation.

Specifying that viewpoint, for us humans, amounts to imposing the image of a clock right in the bivector's confining plane. Bivector and clock thence either go in the *same* direction, or the *opposite* one, unambiguously. Orientation works the same for all expansive numbers:

*Orientation of an expansive number is merely its direction within the smallest space it could inhabit. Within that space, it has only two values; namely the **same** direction relative to something else in there, or the **opposite** direction. Each direction is the negative of the other.*

Bound orientation

Bound elements are already within the smallest space they

could inhabit, so their orientation is most transparent. Of those, the *very* most transparent to us are the elements bound to the same space we are, namely bound trivectors:

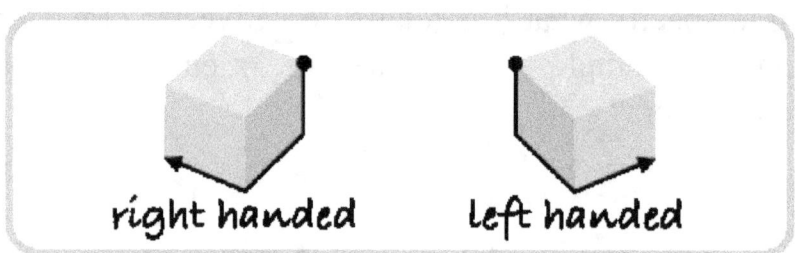

Bound trivectors with opposite orientation.

¿What makes a right-handed trivector right handed? Just this: wrap your hand around the tail segments of its segmented arrow, palm at tail, fingers pointing in arrow direction. Your thumb will point in the direction of the head segment. Similarly for a left-handed trivector. Give it a try.

Negation of a bound trivector reverses the direction of one segment in its segmented arrow. Reversal of the head segment is easiest to understand because that corresponds to reversal of a thumb; but reversal of either of the other two segments also reverses orientation, as you might have fun sketching. You might also have fun reversing one or the other segments of a bound bivector to watch it change orientation too.

Clearly, the thing relative to which we define trivector orientation is our body. Evolution fortuitously equipped us with appendages that have both possible trivector orientations; but that is an extravagance that bivector orientation does not enjoy.

Curiously, we also define bivector orientation relative to our bodies, namely our memory of the image of a clock on the plane of our retinas. As mentioned, we mentally insert that image in the confining plane of the bivector. We can reverse its relative orientation by inserting that image from the other side; but there is another way to reverse bivector orientation: remove it from its confining plane, flip it perpendicular to itself, and reinsert it. This reverses relative orientation by reversing *it*, rather than *us*.

¿Can we do that with a bound trivector? We could if we could leave our space. Then we could view it from the other side, and a right-handed trivector would look left handed.

Even stranger, we could flip ourselves perpendicular to our volume and reinsert ourselves in it. If we survive that maneuver, what had been a right-handed trivector before would now be left-handed *to us*. It is not the trivector that changed, *it is us*. Our friends would suddenly all disagree with us about which is our right hand and which is our left. If we had taken a screw on our adventure, it threads would be reversed. Of course we would be more prudent to perform this maneuver on the screw, or even the trivector, rather than our bodies.

That is the remarkable consequence of flipping perpendicular to a confining space, obvious for those things we can flip, like bound bivectors and bound vectors.

Bound vectors also have orientation relative to our bodies; but that is so obviously arbitrary that we seldom speak of "vector orientation". Rather we say that a vector "points right", "points up", etc. *relative to us*. Nevertheless, this is true orientation because it is *direction within a confining space*, which has only two possibilities, reversed by negation: "points left", "points down", etc.

We have now exhausted all of the bound *expansive* numbers that can be related to the various directions of our bodies; but there remains another kind of numbers that do not have the expanse necessary to relate them to our bodies, namely points. That lacuna makes it impossible to remove a negative sign by spatial reversal; so orientation for points is no longer *relative*, it is *absolute*, given by the sign of their weight.

That sign can be removed from their weight and attached to their location, thereby allowing points to be expressed like all other bound elements, as a *magnitude* times a **directed-locality**. The magnitude is the absolute value of the weight; and the directed-locality is a signed point. This ploy is sometimes conceptually useful, but it is not often algebraically useful—signed-*weight****location** notation is usually more expressive for points.

Free orientation

¿What is the orientation of a free trivector? How could we ever decide?—it has no segmented tail to wrap our hands around; and it has no intrinsic smallest space it could be a direction within.

The obvious answer is to give it those things by extension with a point. ¿But in which direction—to or from? Well, this ploy can clearly be used on all other free elements too, not just trivectors, so we naturally want to get orientation right at the lowest dimension, namely converting a free vector to a bound one. For that we must extend *from* a point to get these vectors pointing in the same direction; yet another reason extension is best expressed *from*.

That stratagem is the one used in the sky-blue figure to relate the expansive free numbers on top to the bound ones on the bottom. (Query: ¿What would an *extension-to* ploy look like?—you might be surprised about the bivectors.)

If you peer at that figure, you may notice something unexpected: Altho there is no intuitive correspondence between the orientations of the free and bound trivectors, there is an intuitive one between the free and bound vectors, which both point left left-up, and the free and bound bivectors, which are both clockwise.

To establish a correspondence between trivectors, we just decided to extend the free one *from* a point. In so doing we notice that it produces a right-handed bound trivector. So we declare that the free one has that orientation too. Strangely, we could also do this the other way: we could notice that the free trivector has counter-clockwise faces when viewed from outside; so we could declare that the bound one has counter-clockwise orientation too.

This is not a whimsy: if you examine the faces of the bound trivector that contain two segments of its segmented arrow, they are both counter-clockwise. This face-ploy can also be extended to a creature living in the plane of a bivector, or in the line of a vector; but I will leave that to my flat-lander and line-lurker readers—we volume-dwellers have no need for such low-dimensional disambiguation.

We do have need for one final distinction. Having decided that points can have *absolute* orientation given by their sign; we must, for consistency, inquire about the orientation of their free parts. This is done in the same way: extend the free part —a signed scalar—from a unit point. We get a signed point. So scalars can also be considered to have absolute orientation given by their sign, like points do. This is perhaps not a very fertile idea. Everything else free has *relative* orientation, a very fertile idea.

Dimension and basis of extensions

The previous chapter developed three kinds of primitive bases: purely bound, anchored free-as-possible, purely free. We saw that the purely bound basis and the anchored free-as-possible one both articulate the same bound space. However, in this chapter we recognized that the latter basis is *much much* more articulate; so that is the basis we shall use henceforth.

By doing so we gain much and lose nothing: The anchor allows us to descend to points wherever we want, simply by translating our origin with free vectors. We may use these points to generate individual bound elements of any dimension, via extension with free elements.

Such bound elements can be manually composed into free elements by adding *separate-but-exactly opposite* ones. This manifests itself in our anchored basis by canceling the anchor, meaning that its weight vanishes.

Conceptually, such free elements preserve their *separate-but-exactly opposite* property amongst themselves; not only under addition via end-canceling, but also under extension via the distributive rule; both of which almost magically regenerate *separate-but-exactly-opposite* ends, as you have seen over and over.

Algebraically, free elements preserve their lack of anchor trivially—it is simply unavailable to free addition or free extension. Hence free things do indeed constitute a sub-algebra of the full geometric algebra.

That sub-algebra is generated by un-anchoring our free-as-possible basis, leaving only the basis free vectors. They establish the *spatial expanse* in which the anchored basis operates. This is a purely free space whose dimension—meaning *degrees of freedom*—of course gets decremented: it is missing the anchor's degree of freedom.

Composite dimension

The previous chapter introduced the lowest possible composite dimension, namely {**1**, **1**}-*without-magnitude*. In abbreviating this dimension as {*1*}, we took the first step toward *coalescing* it into a mere *free* dimensional distinction —a tightly wrapped package that the **algebra** would not unwrap; but that **We** often would in order to understand what the algebra is doing.

This chapter encountered three more composite dimensions having the same numeric value: {**2**, **2**}-*without-magnitude*, abbreviated as {*2*}; {**3**, **3**}-*without-magnitude*, abbreviated as {*3*}; and {**2**, **2**}. That last composite dimension itself *contains* a composite dimension, so it could be written more fully as {**2**, {**2**, **2**}-*without magnitude*}.

That is the last time you will ever see it written that way, however, because we can always unwrap {*2*} in our minds, knowing that the free-as-possible algebra will kindly keep it bundled. We will see the algebra doing even more intricate bundling in even higher spaces in a few chapters, always generated by an …

Anchored free-as-possible basis

The simplest possible basis is the unit scalar 1, dimension {0}. Scaled manipulations of that base have been almost unreasonably effective for about four thousand years in articulating human affairs. This was done from the beginning by giving scalars informal interpretations: apples, oranges, coins, prices, ages, weight, time, on and on and on …

For almost as long, the unit scalar has even been effective as a basis for articulating geometry, despite its lack of geometric locus and dimension. Scaled versions of it were informally interpreted as length, area, volume.

Then about four centuries ago scalars became *wildly* successful in geometry—*collections* of them interpreted as lengths became available. This was such a pivotal moment that I am going to repeat the statement that introduced them:

> *Any problem in geometry can easily be reduced to such terms that a knowledge of the lengths of certain straight lines is sufficient for its construction.* [First sentence in *The Geometry*, 1637, Rene Descartes.]

Despite the huge popularity of this ploy, it has an obvious problem worth restating: scalar lengths do not themselves have any geometric locus or dimension. This problem was patched up historically by a subterfuge—count the number of scalars in the collection and use that number as dimension. This too has been almost unreasonably effective; so effective in fact that its handicaps are seldom recognized.

Here is the main one: it is ad hoc and informal, so **We** must perform all of the dimensional intricacies ourselves. **We** have done that not too badly, perhaps; but you have just seen it done far better, automatically, by a nearly two-centuries-old algebra. The gentle innovator who introduced that algebra described its advantage this way:

> In extension theory there appears a characteristic method of

calculation which transposed into geometry is inexhaustibly fruitful and here (in the theory of space) consists in subjecting spatial structures (points, lines, and so forth) directly to calculation. [*Brief Survey of the Essentials of Extension Theory*, 1845, Hermann Grassmann.]

In order to directly subject points to calculation, we need a point in our algebra; which brings us to our next-simplest basis: (1 *o*), with dimensions {0}, {**1**}. We have just advanced from *scalar* arithmetic to a *scalar-point* arithmetic that articulates a bound **1**-space, a point. Altho this is a point space, it hasn't yet enough spatial expanse to require the weight*distance rule.

Hence, it may not seem very useful, but it is in fact a tremendous advance: that point simply enhances the scalar's four millennia of success—it can add with the scalar and extend with it. When these new formal capabilities augment our old informal interpretations, possibilities become far richer.

For example, (.33 + 107,623*o*) could represent the chance of rain today in Grassmann's hometown of Szezin, plus its population there, viewed as a kind of geometric-algebra origin. Or (4 + 257*o*) could represent the daily cheeseburger intake plus the weight of an American president, viewed as a kind of original, never before seen on this planet. You can easily generate your own original complexities.

As we advance to even more formal capability, there is a tactic that will automatically generate all possible bases, like so: augment each previous base by extending it from a new primitive. Since we have just installed the anchor, every new primitive will be a free vector.

The first one, x, generates this new extended basis: (1, x, *o*,

92

x◂o), dimensions {0}, {*1*}, {**1**}, {**2**}. You see previous base 1 augmented by x◂1, which is just x. Then you see previous base *o* augmented by x◂o. That advances us from *scalar-point* arithmetic to a *scalar-point-vector* arithmetic that articulates a bound **2**-space, a line.

Continuing in this way we can construct the entire extended basis for physical space. Let us list the first few bases to try to discern a pattern.:

1
1 *o*
1 x *o* x◂*o*
1 y x y◂x *o* y◂*o* x◂*o* y◂x◂*o*

Rote exercise: Continue this pattern with the last basis free vector z. It will begin like this: 1, z◂1 (which is just z), y, z◂y, … and so on. Altho this is a mechanical exercise, I urge you to chug thru it because it systematically generates all the extended bases in order; which will shortly be collected into mono-dimensioned bases and reorganized. Best of all, this exercise should suggest that …

A doubling pattern is becoming obvious: each new primitive pairs each existing base with an extension; so it doubles the total number of bases. Hence, a basis that contains n primitives engenders an extended basis containing 2^n bases altogether. Check this from the beginning: (1) contains 0 primitives so it contains 2^0 (meaning 1) base. Then, (1, *o*) has 1 primitive, so it contains 2^1 (meaning 2) bases, and so on.

(A scalar cannot be a primitive because it does not have unit dimension. Scalars, in fact, *derive from* primitives by retraction; similar to the way that line segments derive from primitives by extension, and on up. There is no *on down* below scalars, as the *Retracting* chapter shall explain.)

Other patterns are not so obvious because the augmentation ploy fails to collect together elements of the same numeric dimension; so let us do that ourselves. Since the ploy pushes the initial elements persistently to the right, we have to collect them in reverse to get the natural order. Here they are for the plane, a bound **3**-space.

$$1 \quad \text{zero}$$
$$o, \text{ x, y} \quad \text{one}$$
$$\text{x} \blacktriangleleft o, \text{ y} \blacktriangleleft o, \text{ y} \blacktriangleleft \text{x}, \quad \text{two}$$
$$\text{y} \blacktriangleleft \text{x} \blacktriangleleft o \quad \text{three}$$

Prose numbers indicate numeric dimension, bound or free. You can immediately tell this is a bound **3**-space by looking at its highest-dimensioned element, $\text{y} \blacktriangleleft \text{x} \blacktriangleleft o$. It has **3** primitive extension factors, and the last one is the *o*rigin, meaning it is bound thru there. Here is the same table collected from the extended bases you generated for physical space, a bound **4**-space:

$$1 \quad \text{zero}$$
$$o, \text{ x, y, z} \quad \text{one}$$
$$\text{x} \blacktriangleleft o, \text{ y} \blacktriangleleft o, \text{ z} \blacktriangleleft o, \text{ y} \blacktriangleleft \text{x}, \text{ z} \blacktriangleleft \text{x}, \text{ z} \blacktriangleleft \text{y}, \quad \text{two}$$
$$\text{y} \blacktriangleleft \text{x} \blacktriangleleft o, \text{ z} \blacktriangleleft \text{x} \blacktriangleleft o, \text{ z} \blacktriangleleft \text{y} \blacktriangleleft o, \text{ z} \blacktriangleleft \text{y} \blacktriangleleft \text{x} \quad \text{three}$$
$$\text{z} \blacktriangleleft \text{y} \blacktriangleleft \text{x} \blacktriangleleft o \quad \text{four}$$

Now a pattern is clear: In both of these tables there are the same number of lowest and highest dimensioned bases, namely one; and then the same number of post-lowest and pre-highest; and so until they meet in the middle. Please check.

This correspondence can be expressed algebraically in a surprisingly expressive way: *reflect* an element from the *ceiling*, meaning *the* (single) highest dimensioned element in the extended basis. You will get its *complement*, the element that would extend it to the ceiling—that is the *reflection*.

To clarify this, let us walk thru some complements for a plane, a bound **3**-space whose ceiling is y◂x◂*o*. As always, the unit scalar is the *floor*. The complement of the floor is the ceiling itself because these two elements extend to the ceiling. Conversely, the complement of the ceiling is the floor. The complement of the origin *o* is y◂x — the *free ceiling* — because those two elements extend to the ceiling. Conversely, the complement of the free ceiling is the origin. And so on. (There are nuances about sign that are peripheral here.) If you play with this, you will notice that complements are always expressed using the same number of bases.

To make this pattern more obvious, let us tabulate that number for each of the bound *n*-bases listed so far, low-dimensioned-to-high — *left-to-right*:

$$1 \qquad 0, 2^0 \text{ bases}$$
$$1\ 1 \qquad \mathbf{1}, 2^1 \text{ bases}$$
$$1\ 2\ 1 \qquad \mathbf{2}, 2^2 \text{ bases}$$
$$1\ 3\ 3\ 1 \qquad \mathbf{3}, 2^3 \text{ bases}$$
$$1\ 4\ 6\ 4\ 1 \qquad \mathbf{4}, 2^4 \text{ bases}$$

On top is *the* venerable scalar basis 1. Next is the scalar-point basis containing one scalar and one point; two bases altogether. Then there is the first scalar-point-vector basis, containing one scalar, two primitives — a point and a free vector — and one bound vector; four bases altogether. And so on up to eight and sixteen bases for the bound **3**- and **4**-spaces.

Now a richer pattern is emerging: Each time we extended a row from a new primitive, we generated next-higher-dimensioned bases that augmented existing bases. To see this, look at your rote transition from a plane, bound **3**-space, to physical space, a bound **4**-space. (1+4+6+4+1 = 16)

95

First you extended z from 1. That augmented the 3 primitives listed to its right, namely y, x, *o*, arriving at 4 altogether, listed underneath. Then you extended z from each of those 3 primitives, which augmented the 3 two-dimensioned bases listed to its right, arriving at 6 altogether, again listed underneath. And so on.

Expressed numerically, each succeeding row derives from its preceding row by successive addition of each pair of numbers encountered. This new row is then appended with a 1 at the start and the end. The 1 at the start represents the floor —the unit scalar—and the 1 at the end represents the new ceiling—the previous ceiling extended from the new primitive.

This numeric pattern pervades mathematics, and was called Pascal's triangle after Blaise Pascal, who popularized it. However it was known long before by the Chinese and others. Combinatorial analysis allows you to individually generate the numbers within it using factorials; but the succeeding-row ploy just described is faster, easier and more transparent for our purposes.

The same numeric pattern is generated by a purely free basis. In that pattern, each free ceiling is often called a *pseudo-scalar* because its engagement with retraction gives it scalar-like properties under Clifford's unification. A bound-ceiling's lack of engagement with retraction precludes such properties; a crucial theme in the next chapter, *Retracting*; and a crucial theme in the *Synthesis* chapter too.

Retracting

Doing extension with a point gains *collective* information about argument locations, but loses *specific* information about them. That loss, recall, manifested itself as the confined *freedom of **bondage***, perplexing perhaps. This freedom, we shall now discover, prevents *undoing* this extension from retaining *any* locus information at all—collective or specific. Only *relative* information can be retained.

Retaining such information is nothing more than free-part extraction. This operation, since free is composed of bound, turns out to be the *geometric* essence of *all* retraction, bound and free; so we shall peer closely at it from various vistas. Doing so will generate the *algebraic* essence of retraction. Semantics engenders syntax, you know.

Grassmann began doing extension in two different ways. First, he *informally* extended a point ***to*** a point, which generated a directed line segment—a "*displacement*" he called it—that he mistakenly freed. A *vector*, v, we call it, meaning "*carrier*", echoing Hamilton's ascendent coinage.

Then Grassmann *formally* extended his new vector ***along*** another roving line segment. He had no choice about that: his freed line segments had lost any moorage ***to*** which, or from which anything could be extended.

His *along* ploy gained ascendance; and is still our current one. Altho intuitive, it is inconsistent with his full algebra; a

97

misfortune that effectively froze-out the bound part of that algebra for us, and froze-in the blunder of line-segment rovability.

It caused Grassmann persistent dimensional confusion when he finally began to *formally* extend one point to another, something our geometric algebra community has not begun doing yet.

Happily, his-and-our *along* extension does have a compensating virtue: it is visually evocative. Grassmann promptly *saw* that shape is unimportant: only the *rejection* of v perpendicular to—*equiangular* to—the other vector is relevant; meaning that extension ultimately pays attention only to the part of v *completely outside* the other, denote it $v_\perp$; so he called extension the *outer product*.

This suggested that, for completeness, there must be a product that *projects* v parallel to—*unangular* to—the other; meaning that it pays attention only to the part of v *completely inside* the other, denote it $v_\parallel$; an operation he called the *inner product*.

Inspired by his evocative terminology, I pronounce $v_\perp$ and $v_\parallel$ not in the conventional way as "v-perpendicular" and "v-parallel"; but rather as v-*out* and v-*in*, pithy terms that clearly manifest completeness:

$$v = v\text{-}out + v\text{-}in.$$

Outer versus *inner* was Grassmann's seminal dichotomy for his products. He knew of course that they generate different kinds of things: a parallelogram and a scalar; but he initially assigned dimension only to the parallelogram ("*second order*" arising from a "*first order*" displacement extended with another); but not the scalar.

So he did not begin with a more fundamental dichotomy: whereas his outer product *extends to higher dimension*; his inner product *retracts to lower dimension*. It took him years of innovative meandering to finally emerge with that idea, which he presented in polished form via *supplements* in his second book.

They finally induced him to define the dimension of scalars ("*zeroth order*", *descending from* a first order displacement *retracted* with another); but he never used the word *retract*, the obvious antonym of *extend*. That pithy term would have suggested the most fundamental dichotomy of all:

*In essence retraction **undoes** extension.*

Grassmann eventually recognized this dichotomy too, but only indirectly as a consequence of supplements. Other mathematicians seldom advanced even that far. Hamilton, for example, happened upon a restricted inner product in a purely syntactic way, whose narrow domain obscured its general retractive nature. He tried to read Grassmann, but his close view of his own creation precluded understanding.

When mathematicians began developing matrix algebra, they realized that a row vector times a column vector corresponds syntactically to Grassmann's inner product of vectors (now denoted with Gibbs' dot •); whereas a column vector times a row vector corresponds to his outer product; but the terms *inner* and *outer* were applied only as erudite jargon, having vague geometric significance.

Only Clifford seems to have really understood Grassmann's dichotomies geometrically; altho Peano and Whitehead demonstrated good understanding. These three had all independently reconstructed Grassmann's algebra for themselves, as they testified.

The *doing–undoing* dichotomy, after the meaning of "*in essence*" becomes clear, automatically generates the others. Since we have already been **doing** for an entire chapter; let us now begin **undoing** for another.

In preparation, realize that undoing anything is generally harder than doing it; and I must confess that this chapter was the absolute hardest to write, but the most rewarding—I learned more than I could have imagined. You may have similar experience reading it. My advice when the going gets hard is to keep going until it gets easy, and then come back. You may need several dozen passes—I did. Keep a pencil handy.

To make the going easy at first, let us start in the simplest possible way, namely …

Undoing point extension

We are now exploring. Exploration has the sad property that things can go awry. When they do, it has the happy property that you will have learned how to avoid that situation; unless you died, as Magellan did in the Philippines, and Captain Cook did in Hawaii.

You won't die in this section, but by naively undoing extension improperly at first, you will have learned how to undo it properly at last.

Immediately there is ambiguity: Extension of a generic (underlined) element $\underline{e}$ with a point always produces something bound; but $\underline{e}$ itself could have been either free or bound.

The reason, recall, is that the extension point can be used as the tailpoint for any free vectors in $\underline{e}$. Extension with it thence eliminates every term containing that tailpoint (an annihilating self-extension) leaving only one term—its extension with residual (head) points, a bound element of incremented dimension.

For the most fundamental example, we can ***do*** a bound vector either by extending a point with another point ($\underline{e} = e$ or ***p***), or with a free vector ($\underline{e} = e$ or v). So which does ***undoing*** this extension generate?—point or free vector?

The answer shall be constrained by the just-mentioned completeness of Grassmann's two products, like so: they shall eventually combine into a unified *geometric product*, for which retraction must scale in the same way extension does.

Specifically, scaling a retractor or its retractee must scale its retraction result equivalently; just as scaling an extendor or its extendee scales its extension result. *Respect for scaling*, in better words (automatically induced for these smooth operations by their *respect for summary*, as the previous chapter explained). It has several remarkable consequences.

The most immediate is that, *in the undoing equation*, the retractor must be a *unit*. Naively attempting to undo extension of $\underline{e}$ from weighted point ***pp***, for example, denoted as $(\underline{e} \triangleleft pp) \bullet pp$, could not possibly recover $\underline{e}$, as desired; but would instead generate p^2 times the presumed recovery of $\underline{e}$, owing to double respect for scaling. The only hope of recovering $\underline{e}$ is use a unit point, ***i*** let's denote it:

$$(\underline{e} \triangleleft i) \bullet i \; = \; \underline{e}$$

The switch in notation from ***p*** to ***i*** is intended to emphasize the unit nature of this point. Naked ***p*** could have been used, since it is a unit point itself; but that notation does not generalize smoothly to free-vector retraction, for which a unit vector is denoted as i, by the *looks-like-1* convention.

This unitary notation comes with a double bonus. First, the letter *i* provides a suggestive reminder that we are performing an *inner* product—the retractor point *i* is *inside* its retractee $\underline{e} \triangleleft i$ (as an extension factor of it). Second, this letter frees up *p* for use as a generic *p*oint.

This preliminary attempt at an undoing equation can be transparently verbalized as "$\underline{e}$ *extended from unit point i retracted by i recovers* $\underline{e}$." It does not preclude a scaled retractor—scaling can be applied after the unit retraction. It merely recognizes that undoing extension is distinct from undoing scaling, which would be specified by a different equation: $(\underline{e}s)/s = \underline{e}$, which of course defines *s*calar division.

Enlightening exercise: Show that retractor scaling can be done *before* the unit retraction by pre-absorbing the scalar in generic element $\underline{e}$. In other words, show that $(\underline{e} \triangleleft i) \bullet si = (s\underline{e} \triangleleft i) \bullet i$, which generates $s\underline{e}$ via the undoing equation. This is enlightening because it requires the respect for scaling of not only retraction (to get $s(\underline{e} \triangleleft i)$), but also extension (to get $s\underline{e} \triangleleft i$).

The equation defining retraction—even with its unit enhancement—still isn't quite right because it fails to take extension's *directed* nature into account: ¿How do we undo the reverse extension, $i \triangleleft \underline{e}$? The answer is that the undoing retraction must again be juxtaposed with the doing extension, like so: $i \bullet (i \triangleleft \underline{e}) = \underline{e}$, transparently verbalized as "*unit point i retracting of i extended from* $\underline{e}$ *recovers* $\underline{e}$". This further enhances the extension-undoing equation:

$$(\underline{e} \triangleleft i) \bullet i = \underline{e} = i \bullet (i \triangleleft \underline{e})$$

… which *still* isn't quite right because it gives no indication whether generic $\underline{e}$ must be free $\underline{e}$ (*italicized*) or bound $\underline{\mathbf{e}}$ (**bold**). To reason clearly about this, let us give $\underline{e}$ here dual terminology as the *extendee* in the retractee $\underline{e} \triangleleft i$, and also the *retraction result*. Let us similarly give point i dual terminology as the *extendor* in the retractee, and also the *retractor* of the retractee.

The synonymy of *extendee \ retraction result* and *extendor \ retractor* is crucial for undoing extension in general. For the particular case here, it requires retractee $\underline{e} \triangleleft i$ to be bound because its extendor i is; but, to repeat, the retractee's extendee $\underline{e}$ could be either bound or free, a corresponding ambiguity in the retraction result.

To remove that ambiguity, let us try a bound $\underline{\mathbf{e}}$ first then a free $\underline{e}$ for the simplest possible situation, namely retracting a bound vector $\mathbf{v}$ by a point i *i*nside its confining line. Our mission, in other words, is to discover what $\mathbf{v} \bullet i$ is for two equivalent cases: $\mathbf{v} = \mathbf{p} \triangleleft i$ and $\mathbf{v} = \mathrm{v} \triangleleft i$, where v is the free part of v.

The first case seems obvious: $\mathbf{v} \bullet i = (\mathbf{p} \triangleleft i) \bullet i$, which generates $\mathbf{p}$ via the undoing equation. In words, "*Retracting a bound vector by any unit point on its confining line produces its corresponding headpoint.*"

This is an obvious undoing of $\mathbf{p} \triangleleft i$; and it appears to be well defined because the headpoint of a bound vector becomes unique when its tail is slid over the retractor point. ¿What could possibly be wrong with it?

To find out, apply a little scaling: ¿What is $\mathbf{w} \bullet i$ where $\mathbf{w} = 2\mathbf{v}$? The headpoint of $\mathbf{w}$ is twice as far from i as $\mathbf{p}$ had been; let us call it $\mathbf{q}$, giving $\mathbf{w} = \mathbf{q} \triangleleft i$. Whence $\mathbf{w} \bullet i = (\mathbf{q} \triangleleft i) \bullet i$, which again generates this bound vector's headpoint. We solved that problem.

A new one arises when we apply respect for scaling. We didn't really need to introduce point q; we could have just scaled $\mathbf{v} = p \triangleleft i$ by 2 like so: $\mathbf{w} = 2p \triangleleft i$. Whence $\mathbf{w} \bullet i = (2p \triangleleft i) \bullet i$, which generates point $2p$ via the undoing equation.

Now we have a problem we can't solve: Since extension respects scaling, both $q \triangleleft i$ and $(2p) \triangleleft i$ equal bound vector $\mathbf{w}$, but when we retract these equivalent expressions by i, we arrive at the blatant contradiction that $q = 2p$. *Blatant* because these points have *different* locations and *different* weights.

This problem propagates up to any bound extendee, as will shortly become geometrically obvious; so clearly a bound $\underline{\mathbf{e}}$ will not work in the undoing equation—apparently retraction by a point must produce a free result. ¿But *can* it even do that?—maybe retraction by a point just isn't tenable?

To find out, let us repeat the previous analysis for a free $\underline{e}$, which for this case would be v, the free part of bound $\mathbf{v}$. This gives $\mathbf{v} \bullet i = (v \triangleleft i) \bullet i = v$, from the undoing equation. In other words, point retraction is apparently …

Free-part extraction

To see if that is really true, apply a little scaling just as before: let us discover what $\mathbf{w} \bullet i$ is where $\mathbf{w} = 2v$. Here, we shall need the free part of bound $\mathbf{w}$, which can be expressed either as $w = q{-}i$, or else as $2v = 2(p{-}i)$. The point information here is overkill because we are not allowing ourselves to use any bound elements in the undoing equation; and that makes all the difference:

$$\mathbf{w} \bullet i = (\mathrm{w} \blacktriangleleft i) \bullet i = \mathrm{w} = (2\mathrm{v} \blacktriangleleft i) \bullet i = 2\mathrm{v}$$

Whence w = 2v. That is not a contradiction at all—it is the free identity we began with. But it is very peculiar:

Whoa!—this non-contradiction has the *same form* as the blatant bound contradiction $q = 2p$ did. In other words the free *algebra* here is identical to the bound *algebra*, even tho their *geometry* is completely different—the bound contradiction resides in the *geometric interpretation* of fixed points, not in their *algebraic rules*. That idea shall loom large in the final chapter when we explore how to articulate free with bound.

We have just discovered that retraction with a point loses all the locus information that extension had gained. This is our first foretaste of the fundamental asymmetry between extension and retraction—undoing something turns out to be as problematic in mathematics as it is in life.

The immediate problem here is that doing extension can generate either bound or free, but undoing it can only generate free, a non-obvious result worth trying to make obvious. In preparation, we must resolve some …

Perplexities

¿How does point retraction work with a point outside the retractee's confining space? Is free-part extraction right-retraction *by* a point, or its left-retraction *of*? Does it make a difference? Answers in reverse:

Yes, it makes a difference—retraction has *complementary commuting properties* to extension, like so: a generic primitive *neg-commutes* when **retracted** with even dimension; and *commutes* when retracted with odd dimension, just the opposite of extension.

These properties became inherent in the undoing equation when extension's directed nature was installed. That advance produced an equation sandwiched around the retraction result, whose retractee on each side, namely either $\underline{e} \cdot i$ or $i \cdot \underline{e}$, has *complementary dimensional parity* to its extendor $\underline{e}$, which is the free retraction result.

Hence an *even-parity extendee* commuting with i across the retraction result induces its *odd-parity retractee* to also commute. Similarly, an *odd-parity extendee* neg-commuting induces its *even-parity retractee* to neg-commute too.

This is our first example of the semantics of retraction engendering its syntax; and the most important one too.

Indeed, this converse syntax is so crucial to the entire geometric algebra, bound and free, that it is worth pausing in advance to permanently install it in your mind. Doing so is easy if you focus on the most familiar conventional case, an odd purely free one, like so:

Whereas a free vector *extended* with another *neg-commutes* (think conventional *parallelogram*), that vector *retracted* with the other *commutes* (think conventional *dot-product, scalar product*). This is true of *all* other odd-dimensioned elements when engaged with a primitive. Even-dimensioned elements are just the reverse.

Which prepares us to answer the afore-mentioned not-so-familiar bound case: ¿Is free-part extraction right-retraction *by* a point, or its left-retraction *of*?

Free-part extraction should be right-retraction by a point because that makes a bound vector and its free part point in the same direction, a *very* natural convention.

(It really *is* just a convention, but the price for violating it is perpetual directional confusion when articulating free with bound. Grassmann did violate it owing to his meandering path of discovery—he asserted mid-book that $[\alpha\beta] = \beta-\alpha$, whose reversed order induced confusion about direction. Worse, its ambivalent symbolism induced further confusion about bondage versus freedom, multiplication versus addition, lines versus points ...)

Enlightening exercise: Show that a point's left-retraction of a bound vector would generate a free vector pointing in the opposite direction. (Hint: ¿In the undoing equation, does $i \cdot v$ point in the same direction as v?)

Finally, ¿How does point retraction work with a point p outside the retractee's confining space? It seems impossible— this retractor out-point cannot be made synonymous with an extendor in-point.

However, there is hope: If p could be uniquely decomposed into p-out + p-in, the way v was decomposed into v-out + v-in; then retraction—since it must respect summary like extension does—could deal with each part separately. The hope is that it would just ignore p-out, and deal with p-in as just described for i-in.

That is indeed what shall happen for a generic free vector v, which is typically partially outside, partially inside its retractee. Its decomposition makes v-out *completely* outside, which gets ignored during retraction; and v-in *completely* inside, which is dealt with.

Unfortunately, a point does not decompose in that way—it is never partially outside or partially inside anything. So point retraction simply *does not work* with a point outside the

107

retractee's confining space.

Consequently, "point retraction" has just become an oxymoron: it disallows specifying a point-retractor argument —that point might lie outside the retractee's confining space, an impossible operation. Retraction by a point has just degraded into a *unary* operation, restricted to a single bound argument, which it presumes to retract by a point somewhere within its confining space.

Which induces this question: ¿What weight shall we presume for this presumptuous in-point?—i.e. how shall we scale after the unit retraction? We must presume a unit point, of course, to get free-part extraction. And for that, we must further presume right retraction. That is one restriction and three presumptions too many:

Retraction by a point is just not tenable algebraically.

Nevertheless and happily, the persevering attempt to derive it turns out to have been fruitful—the algebra it generated is crucial for *free-vector* retraction, which *is* tenable algebraically. It obeys the same undoing equation, enhanced a little further; and it obeys all the commuting rules and all the primitive relations just derived, which is all work well done.

Especially well done, since free-vector retraction depends on it, is the restricted operation that point retraction has finally dwindled into.

Algebraically, that operation has become much simpler: free-part extraction. It is trivial in a free-as-possible basis, as the last chapter explained: A bound argument becomes represented as its free part bound from the origin, and then translated from there to its confining space. The translation is irrelevant to the free part, so it gets ignored; and then—in what is left—the origin just gets ignored too.

To illustrate and elucidate, let us walk thru the boundary case, perhaps not so obvious, namely the free part of weighted point *aa*. It would be represented as $a(o+a)$ (under the transparent Cartesian *translate-then-weight* tactic). Translator a gets ignored; and then—in what is left, meaning *ao*—the origin *o* gets ignored too, leaving scalar *a*.

The reason this is not so obvious is that *ao* is not represented as an extension with *o*, as higher-dimensioned bound elements would be. However, we know from the previous chapter that $ao = a \triangleleft o$. Right retracting this by a point in its confining space, namely *o*—the only point in there —gives $(a \triangleleft o) \bullet o$, which generates *a* via the undoing equation.

Finger exercise: Show that *o* *left* retracting of $a \triangleleft o$ also generates *a* because a scalar has even dimension {0}, so commutes with everything under extension.

The just-seen extreme restriction on the retractor point should clarify why point retraction—tho narrowly correct and conceptually illuminating—is not tenable algebraically. Ignoring the origin is much easier.

There is a nuance about that: We are specifically ignoring a *bound-**from*** origin, not a generic *bound-thru* one—the ignored origin should have been expressed at the *tail* in the basis elements.

This is obvious for extracting the free part of an even-dimensioned bound element like **v**: If the basis expresses **v** as $u \triangleleft o$ translated, then ignoring *o* amounts to right retraction by *o*, which generates u = v. If the basis had expressed **v** as $o \triangleleft u$ translated, then ignoring *o* would amount to left retraction by **o**, which generates u = −v.

This nuance is unimportant for odd-dimensioned bound elements like a weighted point because left retraction equals right retraction, as just seen. What *is* important, however, is consistency; so *all* bound basis elements should have been bound *from* the origin before translation.

That is why the *Extending* chapter, when developing an extended free-as-possible basis, began with the origin; and then systematically augmented it by extending existing basis elements ***from*** each new element. That ploy kept the origin at the tail in every element in which it appeared.

Happily, that is the natural convention for ordinary vectors: they have their *tails* anchored at the origin when displayed in a coordinate frame.

This convention has just become unmoored from the formalities owing to the abandonment of point retraction; but it can be reinstalled by denoting free-part extraction formally, and then defining the rules it obeys.

Algebra

My initial impulse long ago was to just append wings so that, for example, the free part of bound vector **v** would be ^**v**^. That turned out to be too cluttered in complex expressions, and needlessly whimsical. A more compact and accurate notation would be $^{[]}$**v** because roving empty brackets constitute the geometric essence of a free element, as shall be elaborated on shortly. This evocative notation is read in prefix order as "*the free part of bound* **v**". Here is the most-important rule it obeys:

$$^{[]}(\underline{e} \blacktriangleleft p) = \underline{e}$$

In words, "*The free part of any generic free element*

extended from any unit point is just that free element." This salvages right-retraction from the just-abandoned point retraction.

Finger exercise: Use this rule, combined with *extension*'s respect for scaling, to show that free-part extraction respects scaling too: $^{[]}(s\underline{\mathbf{e}}) = s(^{[]}\underline{\mathbf{e}})$. Next, use that to show that $^{[]}(-\underline{\mathbf{e}}) = -(^{[]}\underline{\mathbf{e}})$. Finally, use extension's commuting properties to show that $^{[]}(\boldsymbol{p} \cdot \underline{e})$ equals $\underline{e}$ if that generic bound argument is odd; and $-\underline{e}$ if it is even. This appeals to the complementary dimensional parity of $\boldsymbol{p} \cdot \underline{e}$ and $\underline{e}$; so it salvages point-retraction's complementary commuting rules.

The second-and-final rule not only salvages point-retraction's respect for summary, but further generalizes it to generic <u>M</u>ixed <u>N</u>umbers (having arbitrary *<u>free</u>* and \ or **<u>bound</u>** dimensions, possibly composite, or not):

$$^{[]}(\underline{M} + \underline{N}) = {}^{[]}\underline{M} + {}^{[]}\underline{N}$$

Very enlightening exercise: Deploy this equation on a *single* generic free element to show that the free part of anything free vanishes. (Hint: a free element is the sum of *separate-but-opposite* bound elements—¿Are their free parts still *separate*? Still *opposite*?)

This rule's annihilation of a free element makes free-part extraction a universal operation, like extension and addition. If <u>M</u> and <u>N</u> are composite, the rule recurses all the way down to their innermost elements, ignoring the free ones, and extracting the free parts of the bound ones.

So obviously, $^{[][]}\underline{M}$—the free part of a free part—vanishes. (Jargon: this operation is "*nilpotent*", more transparently called *nothing-again*; like an "*idempotent*" operation such as projection would be more transparently called *same-again*.)

The algebraic simplicity of free-part extraction depends on keeping free things bundled—keeping them as free as possible. The geometric complication is that, so tightly wrapped, they might not be understood to be bundles; but rather thought to be monolithic chunks of space like bound elements are.

That misconception would prevent us from acquiring the full geometric algebra; as actually happened going on two centuries now. To finally rectify that historical misfortune, we need vivid imagery that etches into our minds the intrinsically composite nature of free-part ...

Geometry

Geometrically, a free element is a pair of *roving empty brackets*, as mentioned. This may seem peculiar because ordinary brackets nearly always enclose [something], and they virtually never rove *[]* ... *[]*

So, to etch free-element peculiarity into our minds, let us give the idea *needfully* whimsical terminology: **vacant brackets—vackets**, *opposed wings*, able to parallel-soar anywhere; meandering *vac*uous pac*kets*; *vac*ated flapping jac*kets* ... ahem. *Vackets*.

Opposed-point vector vackets, opposed-line bivector vackets, opposed-plane trivector vackets, frolicking parallel to themselves in space—¿Can you see them in your mind? If so, you will never again misconceive them as monoliths, if you ever did.

A vacket gets overburdened and dismembered (to speak memorably) when extended with a point—it becomes weighted down and loses a wing, unable to soar anymore; it becomes a monolith, a shape-shifting one. Take a look:

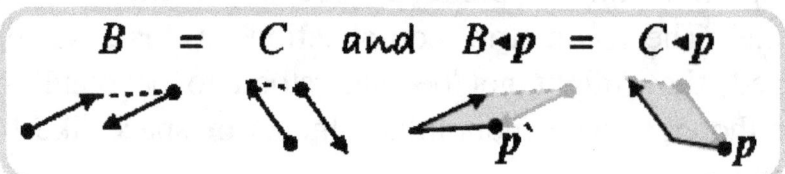

Dismembering and overburdening via point extension.

On the left are two free-bivector vackets, *B* and *C*, cavorting around in space; identical twins in different poses. Here is a memorable story about them:

On the right, free *B* was initially extended from point *p*, which not only overburdened it with the area between its wings, but also dismembered its right wing, whose ghost is shown grayed out.

In other words, free *B* became bound *B*, confined to the plane thru itself, unable to cavort anymore with its ex-twin *C*; who has begun to view it as a distant relative; an *overweight handicapped* one. Bound *B* became lonely; so it slid out of the way to point *p`* and invited free *C* to rejoin it via extension from point *p* too.

C's desire to rejoin beloved ex-*B* overwhelmed its misgivings about *B*'s new condition; so it accepted the invitation and thence became bound *C*: overburdened, dismembered, and trapped in the same plane too.

B and *C* thence became identical again, still with different poses. Sure, they can't *soar* around together anymore, but they can at least *slide* around together; and togetherness trumps everything, even mandatory slithering. Speaking memorably.

Enlightening exercise: Recall the reason free *B* and *C* each lost a wing during extension with *p*: that end had been placed over that point before the extension, which annihilated it. Show that this *poof* method of extension is unnecessary—*B* and *C* can be placed *anywhere*, and each will still coalesce to the same bound bivector, minus a wing, when extended from *p*. (Hint: each will generate two bound (wingless) bivectors that intersect on a line thru *p*. That line can be used as a collecting factor to collect them into one bound bivector. Personal sketches will be crucial—mine were for me.)

Having just completed the previous exercise, you should now realize that there are countless ways to extend anything free with a point. Fortunately, they all arrive at the same result, so it makes sense to take the shortest path there, which is the *poof* method, geometrically.

Algebraically, the shortest path is to stay as free as possible, which keeps free things bundled and tidy; but unfortunately hides their vacated-bracket geometry. Fortunately, there is a different path to bound bivectors *B* and *C* that provides an illuminating alternative view of their geometry, crucial for undoing it: they could have been generated by *p*'s extension with bound vectors, as you see here:

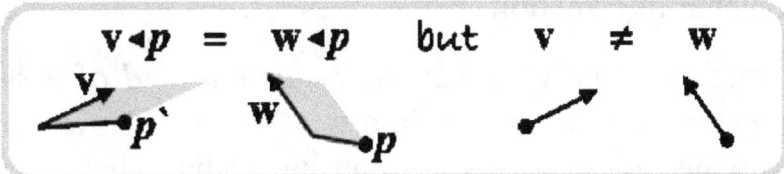

Merely ascending in dimension via point extension.

This figure is illuminating. On the left you see the same bound bivectors as before, *B* and *C*, twins but with different shapes. They were not generated by *p*'s extension with free twins *B* and *C* as before, but rather with bound distant cousins **v** and **w**.

114

Clearly an attempt to retract this B and C by p, denoted as $(v \triangleleft p) \bullet p$ and $(w \triangleleft p) \bullet p$, would generate the contradiction that $\mathbf{v}$ = $\mathbf{w}$. Retraction simply cannot produce a bound result.

This is now *geometrically* obvious, as it had not been before: a bound vector always had a unique headpoint when its tail was slid over the retractor point. Well, *a bound bivector does not have a unique side* when *its* tail is slid over a retractor point. And neither does a bound trivector, and so on up.

The crux of the problem is that the *freedom of **bondage*** allows B and C to transform into each other, but disallows $\mathbf{v}$ and $\mathbf{w}$ to do so. The reason was mentioned at the outset: an extension like $\mathbf{v} \triangleleft p$ gains *collective* (plane) information about its arguments' locations, but loses *specific* (line, point) information about them. That loss is gone forever; only the gain is retained.

Paradoxically, that loss is what retraction must descend into, since it decrements dimension. In the present case, it must descend from plane to line, but **the only information it can retain is planar**. This would be impossible if lines were not able to team up into *separate-but-opposite* pairs, which do retain planar information.

By retaining that information, $\mathbf{v}$—*as one end of a shape-shifting pair*—*can* transform into $\mathbf{w}$. It's magic. However that pair—owing to its exactly opposite nature—has lost the plane's specific locality. That is the geometric reason retraction always produces something free.

It requires us to use more careful language about point extension. The usual way of describing $\mathbf{v} \triangleleft p$ is to say that $\mathbf{v}$ has been *swept* from p back to its original position. If that were the whole story, the sweep would retain information about **v**'s confining line and *p*'s location. But it doesn't.

Algebraically, it acquires a new degree of freedom to explore, but loses memory of how it got there. Geometrically, its segmented arrow acquires a new link along which it can slide. That link is a new tail segment for **v**◂*p*; but it would have been a new head segment for the reverse extension *p*◂**v**. Either link opens up new territory to explore.

So the most informative way to describe point extension is to call it a *sweep-and-link* operation: *sweeping* generates the new space, *linking* provides the freedom to explore it (but only it, which induces the bondage).

To think vividly about this, consider a bound element's segmented arrow to be a peculiar kind of link-jointed snake in which each link is at least a little sideways to all the others. Each sideways direction is new territory in which the snake can slither; but such undulations lose all information about wherein it was launched.

Consequently, undoing this extension cannot be a simple reverse operation: *un-link–and–un-sweep*, which would naively attempt to recover that information. Instead, undoing must be *un-link–bracket–un-fill*. This is a geometric manifestation of the asymmetry between extension and retraction.

Bracketing encloses the space the snake had been linking together. This operation provides the only dimension-decremented information still available about that space. *Un-filling* thence completes the dimension-decrement by vacating the brackets, transforming them into a winged creature, a vacket. Take a look:

116

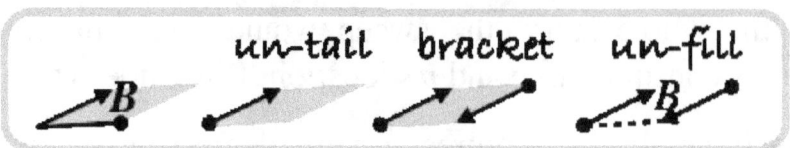

Extracting a free part geometrically

On the left you see bound bivector **B**. The *un-link* operation must be *un-tail* for free-part extraction because that amounts to right retraction at the tail. *Un-head*ing would generate a point's left retraction at the head.

The *bracket* operation simply reverses the end that is left after un-tailing; and uses it to sandwich the space. In this case that end is bound vector **v**, which is reversed and positioned opposite **v** to enclose the space, as you see.

Finally, the area between these opposed ends is *un-fill*ed to complete the descent in dimension, leaving free bivector *B*. This three-step process is how all free parts get extracted, geometrically.

Finger exercise: Try sketching this process on a bound trivector and a bound vector. The slight nuance for the bound vector is that—since its headpoint has no spatial expanse to reverse—reversal becomes simple negation. Next, try the process again on these two elements by un-heading, rather than un-tailing ¿Did you get reversed results? What about un-heading a bound bivector? Reversed result? Why not? Finally, try the three-step process on a weighted point. ¿Is it the *un-tailing* that leaves the scalar weight, or the *un-filling*? (Hint: ¿After the un-tailing, is there any space left to bracket? If not, are the last two steps aborted?)

We have just finished the part of our exploration that went awry—the sad part: We discovered the algebraic futility of undoing point extension, and finally replaced it with simple free-part extraction. We are now prepared for the happy part —exploring the algebraic fertility of …

Undoing free-vector extension

Here is a simple (minded) bound-to-free transliteration of the undoing equation, from a point retractor i to a free-vector retractor, i:

$$(\underline{e} \triangleleft i) \bullet i = \underline{e} = i \bullet (i \triangleleft \underline{e}).$$

Its meaning is that generic element $\underline{e}$ extended from unit free vector i, retracted by i, recovers $\underline{e}$; as does i retracting of i extended from $\underline{e}$.

That meaning is correct; but the equation is a little naive tho it still gets many things right. First, it makes the retractor \ extendor vector a unit for the same reason as before: to preserve retraction's respect for scaling. Second, this unit notation fortuitously suggests again that we are performing an *inner* product because retractor i is *inside* its retractees $\underline{e} \triangleleft i$ or $i \triangleleft \underline{e}$, as an extension factor of them. Third, and most important, it again exposes, with emphatic finality, the impossibility of a …

Bound retractee \ retraction result?

A point retractor i had required a bound retractee $\underline{e} \triangleleft i$, which initially produced an ambiguous free-or-bound result $\underline{e}$. A free-vector retractor i advances to the possibility of a free retractee $\underline{e} \triangleleft i$, which produces the free result $\underline{e}$; but it also seems to permit a bound retractee $\underline{e} \triangleleft i$, producing the bound result $\underline{e}$.

Previously, a bound result had been easily discovered to be

118

untenable for a point retractor; and then a bound retractee was laboriously discovered to be untenable too. Are they *still* untenable for our new free-vector retractor?

Perhaps not, because free-result-from-free-retractee and bound-result-from-bound-retractee do not constitute an ambiguity like free-or-bound-result-from-bound-retractee had. That ambiguity, recall, had been resolved by disallowing a bound result; a disambiguation that does not apply to the current bound retractee.

But that retractee is still susceptible to raw contradiction, which does disallow a bound result. Take a look:

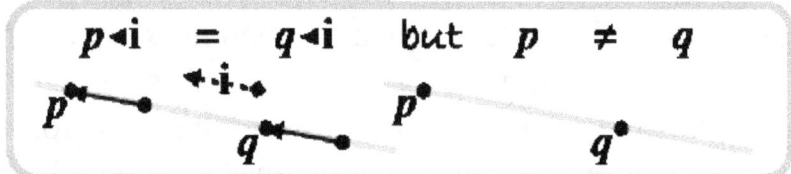

Invalid bound retractee \ extendor \ retraction result.

On the left are two equivalent bound-vector retractees, $p \triangleleft i$ and $q \triangleleft i$, generated by free i's extension to different points on their confining line. Retracting them by i, $(p \triangleleft i) \bullet i$ and $(q \triangleleft i) \bullet i$, would generate the contradiction that $p = q$, via the new undoing equation.

This problem propagates up to all **bound** retractees \ extendors \ retraction results. Consequently, retraction must have *free* arguments only; and it must produce a *free* result. So all vectors, *B*ivectors, Trivectors … hereafter in this chapter shall be free, with no need to say so, except as a reminder:

Retraction is valid in the free sub-algebra only.

Fortunately, you can always enter that algebra via free-part extraction. That operation, we just discovered, regenerates at decremented-and-vacated dimension a bound argument's intrinsic …

Spatial expanse

The intrinsic spatial expanse of our new free-vector retractors raises two new questions that could not have arisen for point retractors: ¿How much expanse? In what direction?

The answer to the first question is unit expanse—*unit separation*—in order to engage the undoing equation; but achieving that is more involved than achieving unit magnitude had been for a point retractor.

Unit weight was easy to achieve simply by translating the origin *o* to the various places where a point was needed, and then declining to scale it, thereby leaving *simple points* having inherent unit magnitude.

Analogous "*simple vectors*" having inherent unit separation do not exist—separation is not an intrinsic and explicit property like weight of a point is. Separation is *extrinsic*—defined elsewhere—and then made *implicit* via that definition.

Defining the separation of a vector requires answering the first question—*¿how much expanse?*—for **every** basis vector in which that vector is expressed. But even that is not enough —the second question must also be answered: *¿What direction* does each expanse have with **every** one of the other basis vectors? Now here is the kicker:

Defining all those basis separations and directions requires retraction. Such a definition is called a …

Metric

The prerequisite for a metric seems to mean that, before we can derive retraction, we must first derive it. Fortunately, the problem is not so chicken-or-egg: before we can derive *generic* retraction, we must first define *primitive* retraction; and that is fairly straightforward.

Defining primitive basis *separation* is especially easy. Say you have decided on a unit of separation: an inch, a meter, a light-year, an anstrom … whatever; and your calculations have arrived at a free vector v whose separation you need to know.

You already know that v can be expressed as vi, where i has your chosen unit of separation; but you don't know what v is yet. ¿How can you find it? Thinking about the undoing equation, you suddenly realize that if vi were expressed as v◄i, you could extract v's separation v via the retraction $(v◄i)•i$.

Unfortunately, you are as ignorant of vector i as you are of scalar v; but you do know that v is their product, vi; so you set up the retraction $(v◄i)•vi = v•v$, and then absorb the retractor's scalar in the extendee v, as previously described: $(v^2◄i)•i$. This generates $v•v = v^2$ via the undoing equation. All that would remain to do is to take the square root to get v's separation v.

Having that, you could define i too, namely v/v. In general, the ability to extract a vector's separation allows you to *normalize* that vector to a unit, crucial for engaging the undoing equation.

However, you still haven't got that separation because v had been expressed in terms of the basis vectors, all of whose separations v ultimately descends to. Here is where the egg begins to generate the chicken—you can just cavalierly *assign* those separations. This establishes the *¿how-much-expanse?* part of the metric.

The only reasonable assignment—if you are designing a general-purpose context-free algebra—is to declare that all basis vectors have a separation of 1. To evoke that value, let's label those vectors as i_1 i_2 i_3, using the *looks-like-1* notation that Clifford preferred. Under that notation, basis separations become defined by the equations $i_j \bullet i_j = 1$ for j from 1 to 3.

These vectors are the free-as-possible part of a bound *o* i_1 i_2 i_3 basis; they are as close to being "*simple*" as vectors can ever get. Their translation of *o* generates simple points i_1 i_2 i_3.

Before v's self-retraction can become completely well defined, we must also establish the *¿what direction?* part of the metric, and that is more tricky. Fortunately, when sign is not given attention, geometry has only two well-defined context-free directions: parallel to, or perpendicular to an element e, more precisely termed *unangular* or *equiangular* to it.

All other directions are ultimately defined in terms of those two angles: a little unangular to e, plus a little equiangular to it. That is the context-free reason right triangles and their trigonometric ratios played such a prominent role in mathematics right from its beginning.

Only recently did such directions achieve direct expression in an algebra, with Grassmann's outer product. It pays full attention to the outer *equiangular* part and ignores the inner *unangular* part. This means that the outer product effectively performs trigonometry for us—*Trigonometry without the trigonometry*, as Heaviside might have said (he called his own algebra "*Quaternions without the quaternions*"). Here is a simple example:

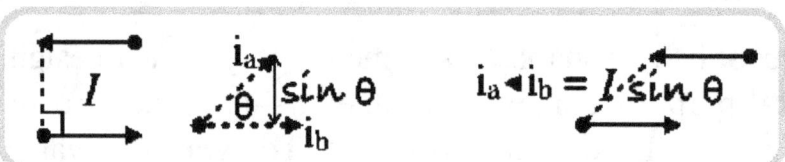

Implicit trigonometry of extension.

On the left is the free ceiling for a plane, a unit bivector *I* formed by extending two perpendicular unit vectors. Had they not been perpendicular, the separation of *I* would have been reduced by the sine of the angle between them, as you see on the right.

Extension's ignoring of the unangular part tells you that part is parallel to $\underline{e}$; but such ignoring pays no attention to sign because $+0 = -0$. Extension's attention to the equiangular part does pay attention to sign, causing $u \blacktriangleleft v_{\|}$ (where $v_{\|} = su$, for some scalar s) to equal $-u \blacktriangleleft v_{\|}$, meaning that it vanishes.

In a complementary way, Grassmann's inner product pays full attention to the inner *unangular* part, including sign; and we shall now discover that it ignores the outer *equiangular* part, neglecting sign. This means that the inner product also effectively performs trigonometry for us, as we shall see.

The primitive case is what we are concerned with now: retraction of two perpendicular vectors, $u \bullet v_{\perp}$. It vanishes because perpendicular projection vanishes; but understanding that as an *undoing* of extension is challenging because its vanishing scalar result is not directly visualizable, like the vanishing area-separation of *doing* parallel extension is.

The best we can do is to conjoin these vectors tail on tail, an operation that ignores vector order like retraction does, and then transform that configuration in a way that always preserves its scalar result. This will transform it into its own negative, showing that the result vanishes. Here is a picture:

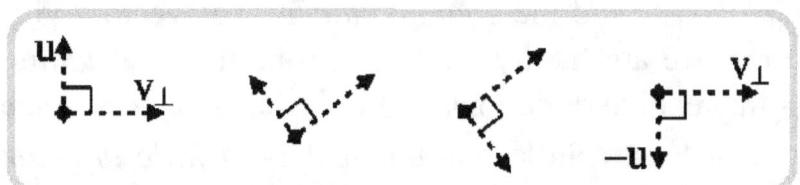

Primitive perpendicular retraction vanishes.

On the left you see two perpendicular vectors, u and $v_\perp$, conjoined tail on tail. Their retraction descends to a scalar, which ignores the absolute directions of the vectors, and pays attention only to their relative *un\equiangular-ness* (the angle between them), and to their separations.

Consequently this configuration can be moved around in any way that preserves those properties. Doing so loses the original vector identities; so those labels have been removed from the transforms, except for the last one on the right. It has suddenly clicked into a position that can be relabeled again, but with a negation:

$$u \bullet v_\perp \; = \; -u \bullet v_\perp$$

… showing algebraically that retraction of two equiangular vectors vanishes, like extension of two unangular vectors does. Notice that this would not be true if the equiangular vectors were not **exactly** so—the transforms could not have clicked into a negated configuration. This is important, so try some non-equiangular sketches if it is not clear.

Also notice that these transforms, altho intuitively obvious for retraction, are invalid for extension because its result ascends to a plane, which does pay attention to absolute directions. In particular, the flip in the middle would have negated that extension.

Now that we know how to precisely express both the *unangular-ness* and the *equiangular-ness* of vectors via their retraction, we are finally ready to define the *un\equiangular-ness*—the relative directions—of the i_j basis vectors. Each one must be at least a little non-unangular—*a little angular*—to all the others for them to be independent; so the question is: ¿How angular?

The only reasonable answer for a context-free algebra is *completely* angular—*equi*angular, *spatially symmetric*, **no unangularness at all**—meaning that $i_j \bullet i_k = 0$ when j does not equal k. When j does equal k, the result should be 1, as just explained. This is called an *orthonormal* basis, meaning *equiangular-unit*.

It is the basis deployed in this introductory book (and most similar ones). Special-context bases can be derived from it, if desired; or else they can be directly specified. It finally makes the extrinsic and implicit separation of v well defined. It prepares us at last to begin developing the simple …

Essence of retraction

Here is the *conceptual* essence: undo an extension of a free-vector retractor that had generated a free retractee element. Such undoing shall generate, in succession, the *geometric*, *computational* and *algebraic* essence of retraction.

Let us denote the *retractor* as free vector r, and the retrac*tee* as free element <u>*tee*</u>. The part of r outside <u>*tee*</u> could not have generated _tee_ by extension; so the first job is to project r inside the retractee, thereby leaving effective retractor $r_{\parallel}$.

To say this algebraically, the other part of r, namely $r_{\perp}$, is perpendicular to _tee_, so that part of the retraction vanishes. This leaves a retractee•retractor in the form _tee_•$r_{\parallel}$.

The next job is to normalize the effective retractor to a unit vector i to preserve retraction's respect for scaling. Since that requires retracting by naked i, the scalar _s_ needed to multiply i to $r_{\parallel}$ should be transferred to the retractee, producing _stee_•i.

Such scalar absorption in the retractee is a wonderful tactic that can abstract away other non-essentials. For example, you may actually want to compute v•_tee_ rather than _tee_•v. To do so, just transpose v•_tee_ into _tee_•v, negating v if _tee_ had even dimensional parity, and then absorb that sign in _s_. Or you may want to normalize _tee_ to a unit, whose normalization factor you can also absorb in _s_. In this way you can focus on _retraction without distraction_.

Finally, we need to express _tee_ as an extension with i in order to engage the undoing equation. This produces $(s\underline{e} \blacktriangleleft i)\bullet i$. Unfortunately, this is naive, as mentioned—free retractee $s\underline{e} \blacktriangleleft i$ is ambiguous in a way that the bound retractee $\underline{e} \blacktriangleleft \boldsymbol{i}$ could not have been, like so:

The spatial expanse of fixed retractor \ extendor i allows $s\underline{e}$ to have an unlimited variety of angles with it. Every different such extendee $s\underline{e}$ would induce a corresponding different retraction result.

The solution is obvious: Since the extension $s\underline{e} \blacktriangleleft i$ pays attention only to the part of i equiangular to $\underline{e}$; require extendee $\underline{e}$ to be _completely_ equiangular to i, denoted as $\underline{e}_{\perp}$.

That will make undoing extension pay *complete* attention to i, as desired. It induces this final ultimately-enhanced vector-undoing equation:

$$(s\underline{e}_\perp \blacktriangleleft i) \bullet i \;=\; s\underline{e}_\perp \;=\; i \bullet (i \blacktriangleleft s\underline{e}_\perp).$$

A person might suppose that the requirement for the extendee to be perpendicular to i would impose such a restriction on the retractee that few things could ever be retracted.

That is not so because that retractee is free, so $\underline{e}$ within it can always be *squared up against* retractor i by shape-shifting, geometrically; or by descent to a basis, computationally; or by symbolic manipulation, algebraically.

Before starting on the path thru those three understandings, pause to fully assimilate the two most important features of the new undoing equation: The result of its retraction is not only *perpendicular* to the retractor vector, as declared by its $_\perp$ subscript; but that result is also *parallel* to its retractee, as an extension factor of it. These are *intrinsic properties* required by extension undoing; not mere peculiarities of this equation.

Indeed, these two properties are so crucial that they are worth restating more precisely and comprehensively: ***every vector*** within the retraction result is perpendicular to the retractor vector; and ***every vector*** in there is also parallel to the retractee element. Every vector.

That is the essential take-away from this chapter if you are traveling light.

Geometry of retraction

Say you have a generic retractee element *tee* that you want to retract by a generic retractor vector r, expressed as *tee*•r. Geometrically, this is a four-step procedure:

- **Project** r onto *tee*:   *tee*•r$_\parallel$
- **Normalize** r$_\parallel$:   *stee*•i
- **Square-up** *stee* against i: $(s\underline{e}_\perp \triangleleft i) \bullet i$
- **Unextend**: $s\underline{e}_\perp$

The most illuminating retraction is that of a *Bivector* retracted by a vector, *B*•r, shown here:

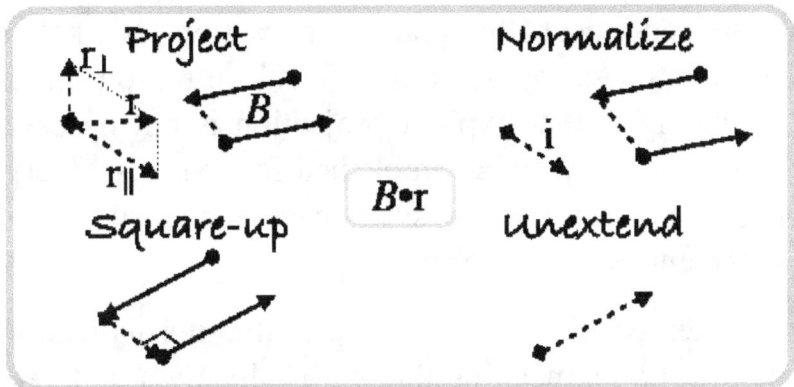

Retracting a bivector by a vector.

On the top left you see bivector *B* being retracted by vector r. The first step is to project r onto *B*, giving an effective retractor r$_\parallel$, as you see. The second step is to normalize that retractor to unit vector i. In this case r$_\parallel$ had to be slightly reduced, compensated by slightly enlarging retractee *B*.

The third step is to square up the enlarged retractee against i. Notice that this step replaces the retractee's dashed-line addition with vector i. To get the sign right, this must be done head on tail as shown.

The final step is to unextend. For this right retraction by a vector, this is always free-part extraction at i's positive point, namely its headpoint. This accommodates the convention that free-part extraction is effectively retraction by a *positive unit point* at the *tail*.

The free part at this end is obvious thanks to the *same-direction* convention for free and bound; but it would still be enlightening to work thru the geometric process: *untail, bracket, unfill*.

It would also be enlightening to undo the undoing by extending the result from i to show that you do indeed regenerate the bivector shown in the square-up step.

The second most-illuminating retraction is that of a trivector T retracted by a vector r, T•r. The reason this is not so illuminating is that explicit projection is not necessary — projection is already an accomplished fact because every free vector inhabits the same physical space a free trivector does, namely the entire available space.

For further simplicity, we may presume normalization has already been done, meaning that r was already a unit vector i, so that T did not need to be trivially rescaled. That leaves this very simple retraction:

Retracting a trivector by a vector.

On the left you see T squared up against i. This is a shape-shifting maneuver that makes T perpendicular to i in such a way that i can replace the dashed-line addition between T's ends, head on tail, as you see.

On the right, this extension was unextended by extracting the free part at i's (positive) headpoint: That bound bivector was untailed, bracketed and unfilled; but those steps may have been so obvious here that you skipped them.

Again, it would be enlightening to undo the undoing to check that you regenerate the free trivector shown in the square-up step.

The least illuminating retraction is the most common one, ironically: retraction of one vector by another. Its result is a scalar, which provides no geometric locus to *square-up* against like other retractions do. Here is a picture:

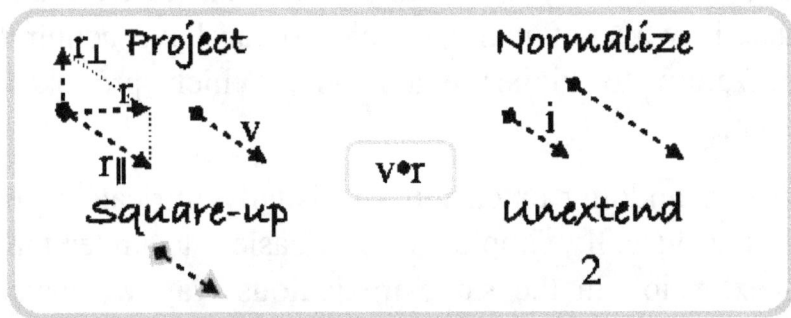

Retracting a vector by a vector.

The project and normalize steps proceed just as before. Again, the effective retractor $r_\parallel$ here had to be to be reduced in order to arrive at unit vector i, compensated by enlarging retractee v, as you see.

Next, the effective retractee, which here happens to be 2i = 2◂i, needs to be "squared-up" against i. This requires its dashed line addition to be resized so that i can replace it, and then the extendee must be reshaped "perpendicular" to that, just as before.

The extendee in this case is a scalar, 2, and scalars are vacuously perpendicular to everything, as we shall see shortly. So the "reshaping" is done in this case by halving the distance between retractee endpoints while doubling their size, shown in gray behind retractor i.

The last step is to unextend 2◂i, namely extract the free part at its positive headpoint, producing 2. A previous exercise indicated that this amounts to simply untailing that weighted point, which leaves something with no spatial expanse available to bracket or unfill, so those steps are aborted.

¿Who knew that the simple dot product of two vectors would be such an intricate operation when performed as an undoing of extension? Not many people so far have known; and that is why not many people so far have acquired its generalization to higher extensions. Which provokes the question ...

¿Why is undoing extension so visually intricate when its doing is so visually simple? This is easiest to understand by *doing* extension in the same meticulous way we have just been *undoing* it, for comparison. The best comparison—since we just meticulously retracted two vectors—is to now meticulously extend them:

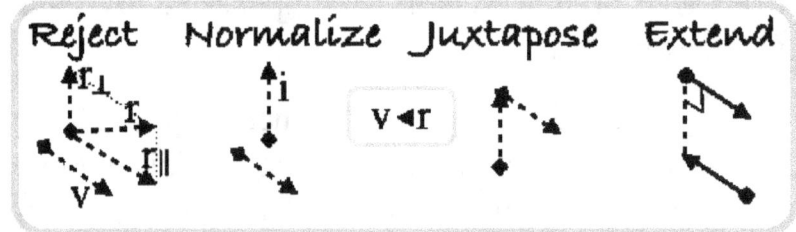

Doing extension as meticulously as undoing it.

On the left you see vector v being extended by vector r, v◂r. The step corresponding to projection of r onto v, $r_{\parallel}$, is rejection of r from v, $r_{\perp}$. This step explicitly recognizes that extension effectively pays attention only to the rejection, and ignores the projection.

However, *explicit rejection is not needed for extension*, visually: we could have gone directly over to the extend step on the far right, v◂r, and produced the same result. True, it would have had a different shape, but that information, tho intuitively appealing, is irrelevant.

Think about what that means for undoing this extension by a presumptuous retractor vector. *Undoing* it requires using information that extension had already *done*, of course; and that is now entirely inside the bivector. Consequently, before we can even start undoing, we must get the retractor vector in there too: **explicit projection is needed for retraction**, visually.

The corresponding next step for extension is to normalize effective extendor $r_\perp$ to unit vector i. For this figure, that requires enlarging $r_\perp$ slightly, thereby reducing extendee v by that amount, as you see.

However, *explicit normalization is not needed for extension*, visually. Again, we could have gone from this step directly over to the extend step, v◂$r_\perp$, and produced the same result, up to shape. This is because *doing* extension *does* its scaling too, automatically, via ascent in dimension.

Consequently, *undoing* extension would automatically *undo* its scaling, a form of division, if its scaling were not *done* explicitly. This idea is so crucial to the distinction between extension and retraction that we it is worth pausing to etch it in our minds forevermore, like so:

Say that you have concocted two simple operations, *ascend* and *descend*, denoted by ↑ and ↓, that take a magnitude (a length, an area … etc.) and raise or lower its dimension by an undirected line segment. (You have temporarily become a Cartesian.) Such undirection, to avoid ambiguity, must be done equiangularly with the magnitude, whatever its direction may be. Here is a picture:

$$\frac{1}{}\uparrow\Big|^2 = 2 = 2 \qquad\qquad 1\downarrow\Big|^2 = 1\downarrow\Big|^2 = \frac{\tfrac{1}{2}}{}$$

Ascent or descent in dimension by a line segment.

On the left, you are *ascend*ing a length of 1 by a line segment of length 2, which generates a rectangle of area 2. Shape and direction are unimportant for this magnitude, so it has been given the most unshaped, undirected area possible, a circle. Dimensional ascent does scaling automatically; meaning that it respects scaling like extension does.

On the right, you are *descend*ing a generic *area* of 1 by the same line segment, for direct comparison. To engage it equiangularly, the area must be formed into a rectangle having one side as the line segment, as you see. The descent is then done by returning the other side, which has a length of 1/2. Consequently, dimensional descent does not *do* scaling; it *undoes* scaling—it *divides*—which certainly does not respect scaling.

That is the conundrum of retraction when it is naively attempted as a simple undoing of extension: its scaling would divide if it were not explicitly normalized. You can try this for yourself: Attempt the retraction steps for *B•*r without normalizing—the unextended result will be *inversely* scaled by the separation of the effective retractor, rather than *directly* scaled by it. So, **explicit normalization is needed for retraction**, visually.

The third step for retraction is to square-up the unit retractor *against* its effective retractee. This has no counterpart for extension because its now-unit extendor, tho now squared-up *with* its effective extendee (via the unneeded explicit rejection), is not inside it; so there is nothing to square-up *against*.

The closest correspondence to *against* is to *juxtapose* the unit extendor onto the effective extendee. However, *juxtaposition is not crucial for extension*, visually. A previous exercise showed that a free extendee can always be extended in place, wherever it fell to earth. Nonetheless, juxtaposition truly is the simplest tactic, visually.

For retraction, however, squaring-up is not merely a tactic: **squaring-up is crucial for retraction**. If it were not performed, the retraction result would be ambiguous.

In summary, whereas extension can be done visually in one simple step; undoing it visually requires four careful steps. So you might worry that *computing* retraction might require much more care than computing extension.

Happily that is not so: Descent to a free-as-possible basis enables retraction and extension to be done with the same amount of care, meaning almost none at all.

Computation of retraction

In a generic free-as-possible basis, vectors u and v would be represented as $u_x x + u_y y + u_z z$ and $v_x x + v_y y + v_z z$. Their retraction u•v would expand into an awful mess, owing to retraction's respect for summary:

$$u_x v_x \; x{\bullet}x \; + \; u_x v_y \; x{\bullet}y \; + \; u_x v_z \; x{\bullet}z \; +$$
$$u_y v_x \; y{\bullet}x \; + \; u_y v_y \; y{\bullet}y \; + \; u_y v_z \; y{\bullet}z \; +$$

134

$$u_z v_x \; z{\bullet}x \;\; + \;\; u_z v_y \; z{\bullet}y \;\; + \;\; u_z v_z \; z{\bullet}z \;\; .$$

With a little care, this could be simplified because retraction of two vectors commutes; so that, for example, $x{\bullet}y = y{\bullet}x$. But such care is not required—that is the boon of descent to a basis: computation can just trundle thru each mindlessly-simple basis operation. Each basis retraction here truly is mindlessly simple because it has been already pre-defined in the metric.

The boon of descent to a basis is also its bane: micro-simplicity is being achieved at the cost of macro-incomprehensibility: it is hard to see the forest for the trees.

Fortunately, in this case you can remove many of the trees to get a fair view of the forest. Simply use an orthonormal basis in which generic x y z would be more transparently denoted as unitary i_1 i_2 i_3. In that form, all the products $i_j{\bullet}i_k$ would vanish for $j \ne k$, and would become 1 for $j = k$, leaving …

$$u{\bullet}v \;\; = \;\; u_1 v_1 \;\; + \;\; u_2 v_2 \;\; + \;\; u_3 v_3$$

… a wonderful simplification. Since this contains nothing but scalar coordinates, a further simplification is suggested: express u and v entirely in terms of those scalars: (u_1, u_2, u_3) and (v_1, v_2, v_3).

This is the modern tuple form of vector notation, pioneered by Hamilton. It gave him just enough view of the geometric forest to discover the crudest kind of meaning for his scalar product, likely achieved like students often do, like this:

Express a retractee vector u of *arbitrary* "length" (*separation*) *u* in a tailored basis for which it has just one basis term. Then retract it by an *arbitrary* vector v with three basis terms, like so (in modern notation): $(u, 0, 0) \bullet (v_1, v_2, v_3)$, which equals uv_1.

Owing to basis orthonormality, this is clearly a projection of v's length onto u, namely v_1, multiplied by u's length, u. To transfer this idea to *entirely arbitrary* vectors, all that remains is to check that coordinate rotation preserves length * projection because it preserves distances (Pythagoras). Consequently, retraction effectively performs trigonometry for us, similar to the way extension had:

$$i_a \bullet i_b = \cos \theta \times 1 = \cos \theta$$

Implicit trigonometry of retraction.

Here you see the same two unit vectors, i_a and i_b, that had generated $I \sin \theta$ by extension. They are now generating $\cos \theta$ by retraction, like so: i_a is projected onto i_b, generating $\cos \theta$ (since i_a's separation is 1). That is multiplied by the separation of i_b, which is also 1, leaving $\cos \theta$.

(Incidentally, when you combine retraction and extension into a juxtaposed product, then $i_a i_b$ becomes $i_a \bullet i_b + i_a \blacktriangleleft i_b$, which reduces to $\cos \theta + I \sin \theta$. Does that look familiar?)

Having discovered the intricate geometry of the scalar product in this way, would you now realize that it could be more generally viewed as generic descent in dimension? Or would you just think that it is merely peculiar, unexpected geometry like its associated vector product?

I certainly *did* think it was just peculiar geometry long ago as a naive student; like all of my classmates and professors. If

136

you had also thought that too, suppose that someone then told you about extension's ascent in dimension.

¿Would you suddenly recognize that the vector product is actually extension in disguise, *without* the dimensional ascent; and that the scalar product is the undoing of extension, *with* the dimensional descent? Not likely.

Hamilton didn't, and he was one of the smartest people who ever lived. He was told about extension and retraction by Grassmann's books and articles; but he did not recognize its expressiveness—his own quaternion baby seemed so much better by comparison.

He did not get even as far as retracting a bivector by a vector, the next most fundamental retraction. It is actually straightforward in an orthonormal basis, as Grassmann knew well. To see that, suppose you have a generic bivector $B = b_1(i_2 \triangleleft i_3) + b_2(i_3 \triangleleft i_1) + b_3(i_1 \triangleleft i_2)$, in cyclic indices. It is retracted by a generic vector $v = v_1 i_1 + v_2 i_2 + v_3 i_3$, expressed as $B \bullet v$.

Direct •multiplication would again generate nine terms—another unfathomable basis computation—but orthonormality again simplifies them. To understand that, begin with the first three terms, which all contain retractor i_1:

$$b_1 v_1 (i_2 \triangleleft i_3) \bullet i_1 + b_2 v_1 (i_3 \triangleleft i_1) \bullet i_1 + b_3 v_1 (i_1 \triangleleft i_2) \bullet i_1$$

The first simplification is that the first term on the left vanishes because i_1 is attempting to retract a basis extension not containing it; which is therefor perpendicular to it (so there is no projection). This is a generalization of the idea that $i_j \bullet i_k$ vanishes if $j \neq k$.

The second simplification is that the second term

immediately engages the undoing equation because orthonormality has effectively *pre-normalized* their retractor \ extendor vectors; and it has effectively *pre-squared-up* that retractee against its retractor, as you see. In consequence, that term reduces to $b_2v_1\,i_3$.

The third simplification is that the third term would also be able to engage the undoing equation if it were neg-commuted to juxtapose the retractor vector against the extendor vector: $-b_3v_1(i_2 \blacktriangleleft i_1)\bullet i_1$, which would reduce to $-b_3v_1\,i_2$. In consequence, the first three terms reduce to ...

$$b_2v_1\,i_3 - b_3v_1\,i_2$$

Let us breeze thru this process again for the next three terms, which all contain retractor i_2:

$$b_1v_2(i_2 \blacktriangleleft i_3)\bullet i_2 + b_2v_2(i_3 \blacktriangleleft i_1)\bullet i_2 + b_3v_2(i_1 \blacktriangleleft i_2)\bullet i_2$$

Inspection shows that the middle term vanishes because $i_3 \blacktriangleleft i_1$ is perpendicular to i_2. Further inspection shows that the right term immediately reduces to $b_3v_2\,i_1$ because it is already in proper form to engage the undoing equation. The left term can be placed in proper form by neg-commuting it: $-b_1v_2(i_3 \blacktriangleleft i_2)\bullet i_2$. Consequently, *these* three terms reduce to ...

$$-b_1v_2\,i_3 + b_3v_2\,i_1$$

Enlightening exercise: Compute the final three terms in the same way to show that the final result of $B \bullet v$ is $i_1(b_3v_2 - b_2v_3) - i_2(b_3v_1 - b_1v_3) + i_3(b_2v_1 - b_1v_2)$. ¿Does that look like a determinate? Indeed, it is the determinate of this matrix:

$$i_1\ i_2\ i_3$$
$$v_1\ v_2\ v_3$$
$$b_1\ b_2\ b_3$$

¿How did determinates get into extension and retraction? It happened in the same way trigonometry got in— automatically, implicitly. Extension and retraction effectively articulate *Determinates without the determinates*.

In fact, it turns out that all properties of determinates— negation on exchange of rows or columns, expansion on a row or column, volume dilation … on and on for multiple treatises in the 1800's—derive from retraction of two same-dimensioned elements, which produces a scalar, unless it vanishes. An enhancement of that process can produce a vector, as just done; or higher dimensioned elements.

We can now compute retraction of a free vector with any free element of the same or higher dimension—just proceed in the same way as done for a free bivector.

What we can't do yet is to compute the retraction of two free elements of arbitrary dimension. To advance to such computations requires the …

Algebra of retraction

There are two ways to proceed: either axiomatically, beginning with the rules and developing their consequences, which are often obscure;   or else conceptually, beginning with crucial ideas from which the rules naturally emerge, and whose consequences are often apparent.

We are already partway down that second path. It began with retraction as the undoing of extension; an idea that quickly generated the following commuting rules: a free vector commutes when retracted with an odd element, and neg-commutes when retracted with an even element; exactly complementary to extension.

To proceed further, we need two crucial squaring-up operations: project one vector *in*side another; and reject one vector *out*side another. They are already conceptually clear, and the *in–out* terminology has already been motivated as pithy expression of Grassmann's distinctions; but now they need to be made effectively computable. That requires some preliminary ...

Primitive operations

Projection and *rejection* require the operations of *separation* and *normalization*. They shall be denoted in the same prefixed-superscript notation as free-part extraction. As with it, special symbols are used to avoid confusion with alphanumeric indexing: $^\$$, intended to evoke $separation; and $^{\prime}$, intended to evoke 1.

- ***vector separation***: $^\$\mathrm{v} = \sqrt{(\mathrm{v} \bullet \mathrm{v})}$: v pithily.
- ***vector normalization***: $^{\prime}\mathrm{v} = \mathrm{v}/\,^\$\mathrm{v}$: i.
- ***vector-vector projection***: $\mathrm{v}\ in\ \mathrm{w} = (\mathrm{v} \bullet\, ^{\prime}\mathrm{w})\,^{\prime}\mathrm{w}$: $\mathrm{v}_{\parallel}$.
- ***vector-vector rejection***: $\mathrm{v}\ out\ \mathrm{w} = \mathrm{v} - \mathrm{v}\ in\ \mathrm{w}$: $\mathrm{v}_{\perp}$.

All of these operations have become computable: Vector separation is just the square root of its self-retraction. Next, a 1-separation vector $^{\prime}\mathrm{v}$—its unit *direction*—is achieved by dividing v by its separation. Then, the projection of v inside w, denoted as $\mathrm{v}_{\parallel}$, is best done by retracting v with w-normalized, times that unit vector; at least conceptually. (It is best done computationally by (v•w/w•w) w.) Finally, since v = v-in + v-out relative to w, the rejection of v outside w, denoted as $\mathrm{v}_{\perp}$, equals $\mathrm{v} - \mathrm{v}_{\parallel}$.

These operations pertain now only to free primitives; but

our journey will move them to generic free elements. The first step is …

Vector retracting a bivector

Say you have a bivector $B = a \blacktriangleleft b$ to be retracted by a vector v: $(a \blacktriangleleft b) \bullet v$. This has already been visualized in a careful way as a four-step unextension; and then computed in an obscure way by descending to an orthonormal basis. We now want to express it in an elegant way by appealing directly to plain vectors a b v.

There is an illuminating tactic for that: decompose a and b temporarily in terms of their projection inside, and rejection outside retractor v. This effectively squares up retractee $a \blacktriangleleft b$ against retractor v, which will expose the bare essence of this retraction. It can then be recomposed back into plain vectors.

The process begins with $((a_\perp + a_\parallel) \blacktriangleleft (b_\perp + b_\parallel)) \bullet v$. It would produce four terms, but $(a_\parallel \blacktriangleleft b_\parallel)$ immediately vanishes because these vectors are each parallel to v, so they are parallel to each other too.

$(a_\perp \blacktriangleleft b_\perp)$ does not immediately vanish: altho these vectors are each perpendicular to v, they are not likely parallel to each other—there are *many* directions perpendicular to v; an entire plane of them in fact, in physical space. However, the *retraction* $(a_\perp \blacktriangleleft b_\perp) \bullet v$ does vanish because $(a_\perp \blacktriangleleft b_\perp)$ is perpendicular to v.

That leaves $(a_\perp \blacktriangleleft b_\parallel) \bullet v + (a_\parallel \blacktriangleleft b_\perp) \bullet v$. The first term on the left can engage the undoing equation if you express v as its separation times its unit direction, vi; and similarly express $b_\parallel$ as $b_\parallel$i (its unit direction is also v's, owing to their parallelity).

141

This generates $(a_\perp \triangleleft i) \bullet i \; b_{\parallel} v$, which immediately reduces to $a_\perp \; b_{\parallel} v$ via the undoing equation. Inelegant $b_{\parallel} v$ expresses b's scalar projection onto v, multiplied times v's scalar separation, all more elegantly expressed as $b \bullet v$. Whence the first term becomes $b \bullet v \; a_\perp$, not quite elegant yet owing to the perpendicular$_\perp$ subscript.

Analogous exercise: By neg-commuting the second term, $(a_{\parallel} \triangleleft b_\perp) \bullet v$, you can square-up its parallel factor against the retractor as just done for the first term. That will engage the undoing equation in exactly the same way, producing $-a \bullet v \; b_\perp$, whence …

$$(a \triangleleft b) \bullet v \; = \; b \bullet v \; a_\perp - \; a \bullet v \; b_\perp$$

This is directly computable via primitive retraction and rejection, and also quite illuminating: It is a sum of vectors perpendicular to v (which are, incidentally, unlikely to be parallel to each other). This demonstrates *algebraically* what we had already discovered geometrically and computationally: converse to the way $\underline{e} \triangleleft v$ *is parallel to* v …

$$\underline{e} \bullet v \textbf{ \textit{ is perpendicular to}} \; v.$$

The price for that information is the inelegant perpendicular$_\perp$ subscripting. ¿Could this equal plain $b \bullet v \; a -$ $a \bullet v \; b$? It could if the parallel part of that expression, namely $b \bullet v \; a_{\parallel} - a \bullet v \; b_{\parallel}$, would just disappear.

It does disappear, and that is easy to see by making it less elegant: $b_{\parallel} v \; a_{\parallel} - a_{\parallel} v \; b_{\parallel}$, an expression that equals $b_{\parallel} v a_{\parallel} \; i - a_{\parallel} v b_{\parallel} \; i$, which vanishes. Whence finally …

$$(a \triangleleft b) \bullet v \; = \; b \bullet v \; a \; - \; a \bullet v \; b \ldots$$

*… **because vector retraction annihilates parallel parts**.*

This means you can always append perpendicular subscripts if you want: $b \bullet v \, a_\perp - a \bullet v \, b_\perp$.

Finger exercise: Repeat this process on $v \bullet (a \blacktriangleleft b)$ to show that it equals $v \bullet a \, b_\perp - v \bullet b \, a_\perp$, which equals $v \bullet a \, b - v \bullet b \, a$.

Retraction of higher-dimensioned elements appeals to the same process, but it becomes more intricate. To elucidate it, let us breeze thru a …

Vector retracting a trivector

The retraction $T \bullet v = (a \blacktriangleleft b \blacktriangleleft c) \bullet v$ starts off as $((a_\perp + a_{\parallel}) \blacktriangleleft (b_\perp + b_{\parallel}) \blacktriangleleft (c_\perp + c_{\parallel})) \bullet v$ when squared-up with v. This would generate eight terms if it were multiplied out, but most of them vanish, like so:

Any term containing more than one parallel factor immediately vanishes. The unique term containing factors all perpendicular to v, $(a_\perp \blacktriangleleft b_\perp \blacktriangleleft c_\perp)$, vanishes as soon as it is retracted by v. Hence, the only terms that survive are those that contain exactly one parallel factor.

There are three places for that factor, so three terms survive: $(a_\perp \blacktriangleleft b_\perp \blacktriangleleft c_{\parallel}) \bullet v + (a_\perp \blacktriangleleft b_{\parallel} \blacktriangleleft c_\perp) \bullet v + (a_{\parallel} \blacktriangleleft b_\perp \blacktriangleleft c_\perp) \bullet v$. The first term on the left already has its parallel factor juxtaposed against retractor v; so it engages the undoing equation as just described, generating $c \bullet v \, (a_\perp \blacktriangleleft b_\perp)$.

The second term can have its parallel factor neg-commuted against the retractor to generate $-b \bullet v \, (a_\perp \blacktriangleleft c_\perp)$; and the third term can have its parallel factor neg-*neg*-commuted against the retractor to generate $a \bullet v \, (b_\perp \blacktriangleleft c_\perp)$. Consequently …

$$(a \blacktriangleleft b \blacktriangleleft c) \bullet v = c \bullet v (a_\perp \blacktriangleleft b_\perp) - b \bullet v (a_\perp \blacktriangleleft c_\perp) + a \bullet v (b_\perp \blacktriangleleft c_\perp)$$

Again, this expression is directly computable via primitive retraction with, and rejection from, v; and its perpendicular subscripts again transparently demonstrate that the result of retraction by a vector is perpendicular to it—*every* *vector* is perpendicular to it.

However, conversion of this expression to plain un-subscripted $c \bullet v(a ◄ b) - b \bullet v(a ◄ c) + a \bullet v(b ◄ c)$ is more intricate because its parallel part is somewhat of a mess (try this at home):

$$c \bullet v(a_{\|} ◄ b_{\perp}) + c \bullet v(a_{\perp} ◄ b_{\|}) -$$
$$b \bullet v(a_{\|} ◄ c_{\perp}) - b \bullet v(a_{\perp} ◄ c_{\|}) +$$
$$a \bullet v(b_{\|} ◄ c_{\perp}) + a \bullet v(b_{\perp} ◄ c_{\|}) \ .$$

Each term here is parallel to v because it contains exactly one factor parallel to v. (But only one—otherwise it would vanish.) It would seem a miracle for all of these terms to somehow magically annihilate each other.

Well, they don't *all* do so; not all at once anyway—they annihilate each other *pairwise* and that is easy to see by descending to scalars as done with bivector retraction:

$$c_{\|}va_{\|}(i ◄ b_{\perp}) + c_{\|}vb_{\|}(a_{\perp} ◄ i) -$$
$$b_{\|}va_{\|}(i ◄ c_{\perp}) - b_{\|}vc_{\|}(a_{\perp} ◄ i) +$$
$$a_{\|}vb_{\|}(i ◄ c_{\perp}) + a_{\|}vc_{\|}(b_{\perp} ◄ i) \ .$$

Notice, for example, that the first and last terms, $c_{\|}va_{\|}(i ◄ b_{\perp})$ and $a_{\|}vc_{\|}(b_{\perp} ◄ i)$, are negatives of each other; and you can check that all the other terms also cancel pairwise.

Consequently, this vector retraction again annihilates parallel parts, allowing you to express $c \bullet v(a_{\perp} ◄ b_{\perp}) - b \bullet v(a_{\perp} ◄ c_{\perp}) + a \bullet v(b_{\perp} ◄ c_{\perp})$ more elegantly as $c \bullet v(a ◄ b) - b \bullet v(a ◄ c) + a \bullet v(b ◄ c)$ if you wish.

Finger exercise: Repeat this process on v•(a◄b◄c) to show that it equals v•a(b$_\perp$◄c$_\perp$) − v•b(a$_\perp$◄c$_\perp$) + v•c(a$_\perp$◄b$_\perp$) which equals v•a(b◄c) − v•b(a◄c) + v•c(a◄b).

To understand the bare essence of this process, let us repeat it one last time, schematically, with a …

Vector retracting a generic element

For transparency, let us perform a vector v = vi retracting of a generic penta-vector: v•(#####), where a hash # in the ith position represents w$_i$; and symbol juxtaposition represents extension. When each # is squared up against retractor v, it generates this alternating sum:

$$s_1(\ \perp\perp\perp\perp) - s_2(\perp\ \perp\perp\perp) + s_3(\perp\perp\ \perp\perp) - …$$

… where a ⊥ in the ith position represents w$_{i\perp}$, and scalar s_i represents v•w$_i$. A blank in the ith position indicates that vector w$_i$ is absent because it has been absorbed in scalar s_i.

You could just stop at this point and proceed to calculate— each s_i, and each ⊥ is a simple primitive operation with v. Indeed, this might be most efficient, depending on how you descend to a basis, because it might avoid needless calculation doomed to subsequently annihilate itself.

However, if you want to express this more elegantly in terms of the plain #s, then you have to show that all of the parallel terms—the ones generated by replacing each ⊥ with its plain #—just disappear.

Each of the perpendicular terms induces as many parallel terms as it has extension factors—one term for each factor parallel to v. The second perpendicular term, −s_2(⊥ ⊥⊥⊥), for

145

example, has $-r_4s_2(\perp \perp i\perp)$ as one of its corresponding parallel terms. The 4th factor is parallel to v, namely $w_{4\|}$, meaning $w_{4\|}$ i, expressed schematically here as r_4 i.

Since there are four factors in each perpendicular term, and five terms altogether, the parallel part therefor has twenty terms; quite a mess to expect to just disappear. But it does disappear because every one of those terms has an anti-term that annihilates it.

The just-mentioned $-r_4s_2(\perp \perp i\perp)$ parallel term, for example, has $-r_2s_4(\perp i\perp \perp)$ as its anti-term. By composing r_4s_2 and r_2s_4 relative to their interaction with retractor v, you will see that they equal each other. Then, by neg-commuting one term into the other, they cancel. To understand that this always happens for each term and its anti-term requires intricate neg-commuting analysis. Have fun.

Or just recognize that vector retraction automatically cleans up its parallel mess; so you can always express it entirely in plain terms, for elegance, if you wish. Or entirely in perpendicular terms, for transparency and possibly more efficient computation.

We have finally become able to retract a free vector algebraically with any free element whatsoever; but there is still one free thing we don't know how to retract it with, namely …

Vector retracting a scalar

A scalar is free, but it is not an element because it can never arise from iterated primitive extension. It can, however, *descend* from iterated primitive *retraction*; and when it does,

146

retraction will either have to deal with it somehow; or else have a discipline for avoiding it.

Both options shall be crucial for retraction's interaction with extension, but they will not become fully apparent until the *Synthesis* chapter. In preparation, let us explore how vector retraction must deal with a scalar—*if it must*—for successful interaction.

Having to deal with a scalar is a real conundrum for retraction because it cannot engage the vector undoing equation—no extension with a vector can possibly *do* a scalar, so there is nothing available to *undo*. That presents a novel dilemma: ¿Who is retracting whom? There are two opposing answers, each with impeccable precedent:

- It is the little guy retracting the big guy. This is what always happened previously in this chapter, just the opposite of what happens in life.
- It is the vector doing the retracting. This is also what always happened previously. However, owing to the fact that *elements* were being retracted, the vector was always the little guy.

Let us go first with the scalar as the retractor because that does not change dimension, which seems most feasible for a scalar operation. It also allows us to resolve the problem of lacking an undoing equation to engage—we can just establish a *scalar* undoing equation:

$$(v \blacktriangleleft 1) \bullet 1 \;=\; v \;=\; 1 \bullet (1 \blacktriangleleft v)$$

This is a perfectly fine, tho awfully simple-minded, undoing equation. Extension of a vector with a scalar has already been validated as scalar multiplication, precisely so it can engage an extension-undoing equation. Moreover, retraction's respect for summary and scaling is being preserved here by a unit retractor \ extendor, namely 1.

However, for this equation to be of any use, we must apply scaling (that is the whole point); either after the unit retraction, or else by absorbing it in extendor v. Either way generates the peculiar result that ...

$$\mathbf{v} \bullet s \;=\; s \mathbf{v} \;=\; s \bullet \mathbf{v}$$

In other words, scalar retraction has become scalar multiplication, just like scalar extension had done in the previous chapter. This gives us pause—retraction is supposed to *undo* extension. Well it can't in this case—that would be division, which fails to respect summary and scaling, as retraction must.

Further pause: Retraction of a vector with an even element like a scalar had been seen to neg-commute; yet it is now commuting. ¿How did that happen?

It happened because the new undoing equation naively gave the retractees, $\mathbf{v} \blacktriangleleft 1$ and $1 \blacktriangleleft \mathbf{v}$, and the extendee \ retraction result v the *same* dimensional parity; which makes scalar retraction commute in the *same* way that scalar extension does, meaning always.

So a scalar as a retractor is not tenable; and our exploratory undoing equation is worthless, as it might have seemed. Retraction and extension are intrinsically *dimension-changing, located* operations that can *engage* a scalar, but not *be engaged by* it—a scalar lacks the dimension and locus needed for that.

So it must be the vector that is doing the retracting, not the scalar (just as it must have been the point doing the extending in $a \cdot a$, not the weight). However, that presents its own problems. It would seem to generate something with a dimension of $\{-1\}$. What would that mean, geometrically?

To answer, we need to carefully rethink the geometric meaning of retraction by v. Its first step, remember, is to project v onto its retractee. ¿So what is the projection of v onto a scalar?

There isn't any of course because a scalar has no locus onto which anything could ever be projected. We already know what happens when v lacks projection: its retraction vanishes. Consequently, a vector retracting a scalar must produce 0.

In other words, a scalar is vacuously perpendicular to a vector. This is a successful interaction: It automatically establishes a scalar floor for geometric algebra—there can be no negative dimensions.

It is also fills a lacuna: retraction of a higher element by a vector has been seen to always generate something perpendicular to the vector. This idea is now extended, er *retracted*, to retraction of a vector by a vector—it produces a scalar, which is perpendicular to the vector.

Then, anything perpendicular to a vector—like a scalar for example—generates 0 when retracted by that vector. The buck stops there—0 is algebraically perpendicular to everything. *Retraction* stops there.

(Summary-respecting products like retraction always generate 0 when engaged with 0: 0 • *Anything* = (*Something* – *Something*) • *Anything* = *Something* • *Anything* – *Something* • *Anything* = 0.)

Such geometric reasoning can be algebraically corroborated: Scalar retraction had been required to commute, or neg-commute, depending on who is retracting whom. Strangely, these rules are not contradictory—scalar retraction can do both: $v \bullet s = s \bullet v$ and $s \bullet v = -v \bullet s$. Combining those rules produces $v \bullet s = -v \bullet s$, whose solution is $v \bullet s = 0$. Similarly for $s \bullet v$.

As seductive these reasonings may be, they do not *prove* that scalar retraction vanishes. That is a new assumption, based ultimately on geometric meaning, like the old assumption that extension of two points neg-commutes. The fundamental test of this new rule is how well it interacts with that old rule, and others, to express geometric meaning. We will keep an eye out.

But not for long because we are heading into the home stretch, beginning with …

Same-dimensioned retraction

Extending free elements ultimately proceeds one vector at a time; so you might expect retracting them to do the same. Indeed, that is one successful tactic. The clearest way to demonstrate it is to begin as simple as possible, with self-retractions of non-primitive elements. Once we master that, we can advance to arbitrary same-dimensioned retraction; and onward to differently-dimensioned retraction.

The simplest possible non-primitive elements are bivectors; and their simplest possible representation is a scalar s times a unit planar direction I. I's simplest possible extension is $i_a \blacktriangleleft i_b$, where i_a and i_b are unit-vector directions perpendicular to each other. In other words, we are beginning with bivector …

$$B \;=\; sI \;=\; s(i_a \blacktriangleleft i_b)$$

We are going to retract this bivector with itself to see how iterated retraction works. For that purpose, there are two nice features of its simple symbolism. First, it is pre-normalized and pre-squared-up, which will immediately engage the undoing equation.

Second, symbol s, like the symbols i and i in the undoing equation, has suggestive dual meaning: s is a *scalar*; and it is *separation* too. This shall be important.

Before beginning iterative retraction, recall how iterative extension works: $B = ((s) \blacktriangleleft i_a) \blacktriangleleft i_b$. In words, B arose by extending scalar s from i_a, and then extending that result from i_b.

The result of *that* is going to get retracted by B, so we would naturally expect it to undo B, except for a scalar, s^2, owing to double respect for scaling. (In the same way that $v \bullet v = vi \bullet vi$ undoes v except for scalar v^2, again owing to double respect for scaling.)

Let us see if it does: $B \bullet B = sB \bullet (i_a \blacktriangleleft i_b) = s(B \bullet i_a) \bullet i_b$. We are simply retracting, one after the other, by each of the extension vectors in the retractor (where the inessential scalar s has been placed outside the essential iteration). To engage the undoing equation, we need to express the retractee in terms of its unit vectors also: $s^2((i_a \blacktriangleleft i_b) \bullet i_a) \bullet i_b$ (with *its* scalar also placed outside the iteration).

Now we have a conundrum—we can't proceed with the undoing equation unless we reverse the retractor (or retractee) by neg-commuting it: $-s^2((i_a \blacktriangleleft i_b) \bullet i_b) \bullet i_a$. This engages the undoing equation twice, leaving $-s^2$. So we did not *undo* bivector I, we *neg-undid* it.

Then it hits us: to undo an extension, we must of course undo its vectors in reverse order of their extension. That makes us think: ¿Maybe retraction should just implicitly reverse its retractor to ensure undoing?

Unfortunately, that would not allow retraction to team up with extension as a unified product. That product requires all extension factors to remain in their original order, with no implicit rearrangement, as you might expect.

However *explicit* rearrangement, namely reversal, can always be used to ensure undoing of a bivector, up to a scalar. Doing so ensures that that scalar is positive (since it is a square), so its square root can be taken to extract separation s.

Then, with bivector separation available, its unit direction I can be extracted by division. These ideas apply to self-retraction of any higher-dimensioned element, as is straightforward to check. Consequently, the first two primitive operations, separation and normalization, can now be generalized to generic free elements of any dimension.

Generalizing the next two primitive operations to generic elements depends on how you generalize vector projection onto a vector. Probably the most obvious way is vector projection onto a *bivector*, and on up. Unfortunately, the disparate dimensions there make that an intricate operation we will not be able to handle until the end of this chapter.

The less obvious generalization of projection is *bivector* onto a bivector, and on up. Let us call this *n-n* projection, meaning that one *n*-vector is projected inside another *n*-vector. It is straightforward because its result is a scaled version of the projectee's unit direction; and that scalar is easy to extract via retraction.

Doing so requires advance from self retraction to mere same-dimensioned retraction. The simplest possible one is bivector B retracting a different bivector, C let's call it.

The simplest different bivector C is one with the same direction I, but a different scalar r. This is still a self retraction: $B \bullet C = sI \bullet rI = rs\, I \bullet I$, which equals $-rs$. The slight novelty is that scalar $-rs$ might not be negative because r and s might have opposite signs (meaning that B and C have opposite orientation).

Happily, same-dimensioned retractions do not get any more complicated than that—even completely arbitrary ones always descend to individually-scaled self retractions, unless they vanish. This becomes algebraically obvious for bivectors when expressed in an orthonormal basis:

$$B \;=\; b_1\, i_2 \blacktriangleleft i_3 \;+\; b_2\, i_3 \blacktriangleleft i_1 \;+\; b_3\, i_1 \blacktriangleleft i_2$$
$$C \;=\; c_1\, i_2 \blacktriangleleft i_3 \;+\; c_2\, i_3 \blacktriangleleft i_1 \;+\; c_3\, i_1 \blacktriangleleft i_2.$$

Their retraction, $B \bullet C$, would generate a customary nine-term mess, except that all cross-terms vanish because they are perpendicular to each other, just as happened with vector retraction.

Consider, for example, the cross-term $(b_1\, i_2 \blacktriangleleft i_3) \bullet (c_2\, i_3 \blacktriangleleft i_1)$. It equals $b_1 c_2\, (i_2 \blacktriangleleft i_3) \bullet (i_3 \blacktriangleleft i_1)$, which expands into the iterated retraction $b_1 c_2\, (i_2 \blacktriangleleft i_3) \bullet i_3) \bullet i_1)$. The undoing equation collapses the innermost retraction, leaving $b_1 c_2\, (i_2 \bullet i_1)$. The final iteration, $i_2 \bullet i_1$, vanishes because i_2 is perpendicular to i_1; so this cross term vanishes.

All other cross terms are similarly perpendicular; so they all similarly vanish. The surviving retractions are individually-scaled self retractions:

$$B \bullet C \ = \ b_1 c_1 \, I_1 \bullet I_1 \ + \ b_2 c_2 \, I_2 \bullet I_2 \ + \ b_3 c_3 \, I_3 \bullet I_3$$

… where $I_1 = i_2 \triangleleft i_3$, and similarly for the other I's in cyclic indices. Since their self-retractions are all negative, $B \bullet C$ becomes …

$$-b_1 c_1 \ - \ b_2 c_2 \ - \ b_3 c_3$$

…which is nothing more than a negated version of a corresponding vector•vector retraction.

(Indeed, this is precisely the form of Hamilton's vector•vector retraction. If he had been able to advance to the bivectors that Grassmann's book had told him about, he likely would have seen that his vectors are actually standing in for bivectors; a ploy that only works in physical space where they can be orthogonal complements of each other.)

Notice that the $b_i c_i$ are merely the products of mutually perpendicular projections onto the Cartesian coordinate *planes* (but negated); just as the previous $u_i v_i$ were the products of mutually perpendicular projections onto the coordinate *axes*. Unfortunately, such projection-*products* are geometrically obscure, as already seen with vector-vector retraction.

Fortunately, they can be made as transparent as that had been; and in the same way: by using a retractee-tailored basis for which retractee B has just one basis term, rather than three. Take a look:

$$B = 3(i_1 \triangleleft i_2)$$
$$C = 2(i_1 \triangleleft i_2) + i_2 \triangleleft i_3$$
$$B \cdot C = 6(i_1 \triangleleft i_2) \cdot (i_1 \triangleleft i_2) = -6$$

Same-dimensioned retraction is a projection-product.

Bivector retractee B here has been expressed as simply as possible in just one basis extension, namely $3(i_1 \triangleleft i_2)$. Retractor C is an arbitrary bivector, which could have been expressed in three terms. However, two terms—a completely overlapping one, $2(i_1 \triangleleft i_2)$, plus a partially overlapping one, $i_2 \triangleleft i_3$—suffice to illustrate the crucial features of this retraction,

… which is that only the completely overlapping term—the *parallel* one—is paid attention to. Partially overlapping terms, or non-overlapping ones, are perpendicular to the retractee to various degrees; so they get ignored during the retraction.

For the case at hand, C's $(i_2 \triangleleft i_3)$ term only partially overlaps the retractee $3(i_1 \triangleleft i_2)$, so it iterates down to two different basis vectors, i_1 and i_3, whose retraction vanishes. C's $2(i_1 \triangleleft i_2)$ term completely overlaps the retractee, so it iterates down on identical basis vectors, $6\, I_3 \cdot I_3$, which produces -6. That is the algebra.

Here is the geometry: C's two terms place it one area-separation above retractee B, and two area-separations over it, as you see. The *above* part is perpendicular to the retractee, so it gets ignored.

The *over* part is effectively projected onto the retractee, as shown by the light gray arrow. Their self-product is the result of the retraction. Such a projection-product parallels vector•vector retraction, except for the implicit negation.

That negation was required so that the retractor could undo the retractee's vectors in reverse order of their extension. If we make reversal an *explicit* operation, then there will be no need for implicit negation in any same-dimensioned retraction; and we can finally convert all previous primitive operations into …

Generic operations

- *reversal*: $\overset{\leftharpoondown}{}(v_1^\blacktriangleleft \ \dots \ ^\blacktriangleleft v_n) = (v_n^\blacktriangleleft \ \dots \ ^\blacktriangleleft v_1)$
- *separation*: $^\$\underline{e} = \sqrt{(\underline{e} \bullet {}^\leftharpoondown \underline{e})}$: $s, v, b, \dots$
- *magnitude*: $^{\$[]}\underline{e}$
- *normalization*: $'\underline{e} = \underline{e}/\,^\$\underline{e}$: $i, I, \mathrm{I}, \underline{i}$
- *n-n projection*: $\underline{n\text{-}e}$ in $\underline{n\text{-}f} = (\underline{n\text{-}e} \bullet {}'\underline{n\text{-}f})\,'\underline{n\text{-}f}$
- *n-n rejection*: $\underline{n\text{-}e}$ out $\underline{n\text{-}f} = \underline{n\text{-}e} - \underline{n\text{-}e}$ in $\underline{n\text{-}f}$

The three novelties here are the reversal needed to ensure uniform positive undoing, rather than variously-negated undoing; the use of generic free n-elements $\underline{e}$, $\underline{n\text{-}e}$, $\underline{n\text{-}f}$, rather than primitives v; and the use of generic element $\underline{e}$, free or bound, in the expression of magnitude.

Magnitude is defined as the separation of a free part; which makes it as universal as free-part extraction; even tho separation itself is only valid in the free sub-algebra. This definition ensures that the magnitude of anything free vanishes.

Visualizing n-n projection gets hard beyond $n=1$. Bivector-bivector projection is probably as far as we low-dimensioned creatures can visualize; and bivector-bivector rejection may be even harder, if only because it is unfamiliar. Visualizing higher-dimensioned projections and rejections might be impossible because we can't reject ourselves from physical space.

The same-dimensioned retractions in these operations all generate a scalar if they do not vanish. This is obvious because descending from dimension $\{n\}$ by that same dimension generates dimension $\{0\}$.

What might not be so obvious is that same-dimensioned retractions pay no attention to argument order. The reason is that a retraction having identical arguments obviously commutes; and since commuting properties depend only on dimension, therefor identically-dimensioned arguments commute too, regardless of whether they themselves are identical or not.

Consequently, who is retracting whom is irrelevant for such arguments, meaning that iteration can be done in either order, from left to right or right to left; and by either argument in either order.

This is perfectly valid macro reasoning; but it may not convince readers who prefer micro detail. The next section provides such detail, and also reveals how atypical this dual indifference to argument precedence is.

Differently-dimensioned retraction

For such arguments, the first question is ¿*Who is retracting whom*? and the second question is ¿*How is it doing it*? Historically there have been almost as many different answers as there are different possibilities.

For the *How?* question, there are two broad possibilities: either primitive-by-primitive in an open-ended way; or else all at once by a kind of dimensioned reflection from an explicit ambient space, in a closed way.

Primitive-by-primitive retraction meshes well with primitive-by-primitive extension; and it does appeal to an ambient space, *implicitly* via whatever basis happens to be in use. These properties establish its open-ended character, and allow it to merge with extension in an especially simple way, done in the *Synthesis* chapter. For these reasons, *primitive-by-primitive* is the answer this book makes to the *How?* question.

The *Who?* question is more nuanced; so let us peek at some historical answers first. Say you have a bivector $B = 3\ v \triangleleft w$, retracted with a trivector $T = x \triangleleft y \triangleleft z$, expressed in that order. There are two ways to iterate, either from left to right, or from right to left.

Let us try left to right first. It is a directed kind of retraction, so it merits a directed retraction symbol, $B \gg T$, let's say, which equals $3(v \bullet (w \bullet (x \triangleleft y \triangleleft z)))$, retracting vector by vector, w by v.

(Recall that the directed nature of extension had also merited a directed symbol; but the difference is that retraction's direction is *algebraic*, denoting direction of iteration; whereas extension's direction is *geometric*, denoting direction of sweeping. For extension, direction of iteration is irrelevant—either direction produces the same result, a manifestation of its associative law. In other words, who-is-extending-whom is irrelevant for extension. That is why you may always read an extension backward; or indeed, backward-and-forward if you want (provided you carefully preserve its direction of sweeping, of course).)

The inner retraction w•(x◂y◂z) produces a bivector: w•x(y◂z) − w•y(x◂z);   and then the outer retraction together with scaling produce a vector:

$$3(v•y\ w•x\ z − v•z\ w•x\ y − v•x\ w•y\ z + v•z\ w•y\ x).$$

Intrepid readers might have fun deriving this; but it is somewhat of a mess even in its plain-vector form. It would be virtually incomprehensible page-long hieroglyphics at the basis level where computation is actually done; and probably no one would have fun deriving that (tho you have become equipped to do so). Nevertheless, even without doing so, we know a surprising amount about it:

We know that the first retraction had produced a bivector perpendicular to w—*every* *vector* in that plane is perpendicular to w—and then, **within** *that plane*, the second retraction had produced a vector perpendicular to v. Consequently the final vector, u let's call it, is perpendicular to both v and w, to v◂w. We don't know in which direction u is perpendicular to v◂w, or what its separation is; but we do know that separation is scaled by 3.

¿See how much the semantics of retraction can tell us about its geometry, without engaging the syntax of its algebra?

Knowing all of that, let us try right to left retraction, denoted as *B*«T, which equals (((v◂w)•x)•y)•z, again retracting vector by vector, x by y by z. Without descending into the gory details we already know, the innermost retraction produces a vector, the next retraction produces a scalar; so the final retraction produces zero, which tells us virtually nothing about the original arguments.

Nonetheless, the final retraction is a useful and valid scalar retraction because it prevents dimension from descending below the scalar floor. The overrunning retraction that arrived at it may seem useless because it provides no information about geometric relations.

It doesn't even allow you to infer that the left argument must have been the little one—the zero result could have also arisen if the other argument had been the retractor, provided it was perpendicular to the big one. Nevertheless, algebraically-directed kinds of retraction like this have been historically popular.

Before pondering why, take a look at a useless and invalid scalar retraction generated by "enhancing" the first retraction: $(3\blacktriangleleft v\blacktriangleleft w)\gg(x\blacktriangleleft y\blacktriangleleft z)$. This seems more elegant because its bivector scaling has been uniformly expressed in terms of extension, giving the arguments a nice visual symmetry.

We already know that this should produce a vector $u = v\bullet(w\bullet(x\blacktriangleleft y\blacktriangleleft z)$, scaled by 3; but in this case the scalar doesn't scale; it just keeps on retracting, $3\bullet u$, and produces zero, which is not valid.

Actually, it is the *vector* u that got control and did the retracting. If the *scalar* 3 had been allowed to proceed, it would have just scaled that vector, as we saw in a previous section; and that would have been valid. The distinction is that $3\gg u$ would be multiple-addition $v+v+v$; but $3\ll u$ would be dimensional descent, a distinction the undirected dot • failed to enforce.

The remedy is to always use juxtaposed multiple-addition, 3u, for scaling, except when dimensional ascent is required for the explicit purpose of undoing it. That way, when scalar retraction arises in a computation, it is probably a valid

160

attempt to descend in dimension, rather than to perform multiple-addition; so it should get blocked by annihilation.

Actually, there is a better remedy: avoid scalar retraction. Doing so requires a discipline acquired by first understanding why its use in directed retraction has been historically popular.

Somewhat superficially, it is because either direction of retraction can be converted into the other by using dimensioned neg\pos-commuting rules. This preserves who is doing the retracting; so it is uninformative when that happens to be the big guy. When that is the little guy, it does inform about how orientation is affected by exchanging arguments.

Also superficially, when programming, directed retraction is sometimes useful in a loop to consistently retract by a fixed retractor, and to return zero if it happens to overrun its current retractee.

However the main reason for its initial popularity was the view that the inner product is merely another kind of product on par with the outer one. That product was known to be a directed one that returns zero when it hits the ceiling; so it seemed obvious that the inner product would be a directed product that returns zero when it hits the floor.

What could be more natural?

This view arose from a poor understanding of the inner \ outer dichotomy, for which *inner* and *outer* had lost most of Grassmann's precise geometric meaning. When this weakened dichotomy advances to a sharper dimensional descent \ ascent one, the view changes. Descent \ ascent arises from dimensional subtraction \ addition, which does not have the appealing symmetry that inner \ outer had seemed to have.

161

Specifically, addition of dimension is open-ended and associative; but subtraction of dimension is closed and non-associative: 7–(3–5) does not equal (7–3)–5. That helps clarify why the inner product lacks the outer product's associativity; but it does not clarify why subtraction of dimension is closed—¿Why are there no negative dimensions?

Historically, there had been no negative numbers either, for thousands of years. However, one of the great triumphs of mathematics was the introduction of negative numbers consistent with positive ones. So I am guessing there must have been determined efforts to similarly concoct negative dimensions too, consistent with positive ones. I personally tried for a few years, long ago and far away.

I found no way to make them geometrically meaningful. What was meaningful is for dimensional descent to just throw up its hands and say *"Fine, you have finally descended to nothing at all. The only direction from here is up."*

The great beauty of the doing \ undoing dichotomy (aside from its automatic inclusion of the other two) is that it does not ever require retraction to throw up its hands like that—it shields retraction from having to deal with an undimensioned retractee, as we shall see.

Another nice feature is that it disciplines retraction in precisely the way needed to combine it with extension. Or, to say this conversely, their combined product automatically induces a doing \ undoing dichotomy. The *Synthesis* chapter will explore this; but for now let us explore how that dichotomy answers …

¿Who is retracting whom?

The little element is always retracting the big one, for this reason: Undoing extension requires something already done to undo. Whatever had been done lies somewhere within the retractee, meaning that it is the same size as, or smaller than the retractee.

Specifically, in the previous retraction $B \bullet T$, the bivector B must be the retractor, meaning that $B \bullet T = v \bullet (w \bullet (x \triangleleft y \triangleleft z))$. Moreover, even in the commuted retraction $T \bullet B$, the bivector must still be the retractor: $T \bullet B = ((x \triangleleft y \triangleleft z) \bullet v) \bullet w$. (Query: ¿Do these have the same signs, or opposite signs? Answer coming up.)

(So you see, an undirected symbol like Gibbs' dot $\bullet$ is well suited for retraction, which is bidirected. A bidirected symbol like $\Leftrightarrow$ or might have been even better. Or not—these symbols are harder to write, even tho more evocative. (But if you like one, use it, and let survival-of-the-most-appealing determine whether it survives.) It is unlikely that Gibbs gave this much thought: his retraction never dealt with non-vectors; and his undirected $\times$ denotes a directed operation, the so-called vector cross product.)

The worst that can happen under a little-element retractor (leaving aside perpendicularity) is that it might just annihilate dimension, producing a scalar.

The only way that could ever happen is if the big element were the same size as the little one, in which case there would be no argument residue left to continue the retraction, which would promptly halt. And *that* is how little-element retraction automatically avoids a scalar retractee.

(Of course, it is always possible to explicitly confront

163

retraction with a scalar; like it is always possible to confront division with a zero divisor: $a \bullet v$, $(u \bullet v) \bullet w$, ... on and on. When retraction encounters a scalar, the algebra will need a strategy for dealing with it. Stay tuned.)

Farewell undoing equation

The previous extension-undoing equations, except for the short-lived and invalid scalar one, all attempted to undo extension by a *primitive*. The attempt to undo extension by the most primitive primitive—a point—failed until its algebra was severely narrowed to mere free-part extraction. The attempt to undo extension by a free vector failed when its extendee \ retractee were bound, but succeeded when they were free, narrowing retraction to the free sub-algebra.

Finally, attempts to undo extension by a generic free element just now succeeded by using iterated primitive retraction. That is the end of the story for *computing* retraction, now that we understand that the little element is always doing the retracting. But it is not the end of the story for *understanding* retraction.

For example, except for *n-n* projection, we still don't understand how to express retraction's projective properties. And we don't quite understand how to articulate retraction's perpendicularity in a way that complements extension's parallelity.

Finally, we still don't understand retraction's commuting properties. Tedious iteration could *tell* us, but would fail to give us understanding. For that we need to retract by an element in its entirety.

Until now, an element in its entirety could not have been

164

used as a retractor in an undoing equation because it might have been too large to serve synonymously as the extendor in there—it might not have fit into its retractee, leaving nothing available to undo.

The fresh requirement for the small element to be the retractor changes all of that—such an element can serve synonymously as the extendor; such an element fits into its retractee. Take a look:

$$(s\underline{e}_\perp \blacktriangleleft i) \bullet \overset{\curvearrowleft}{\underline{i}} \ = \ s\underline{e}_\perp \ = \ \overset{\curvearrowleft}{\underline{i}} \bullet (\underline{i} \blacktriangleleft s\underline{e}_\perp).$$

This is a slightly tweaked version of extension's *"final ultimately-enhanced **vector**-undoing equation"*. Its most slight tweak is that unit extendor *vector* i, of dimension {1}, has been replaced by unit extendor *element* $\underline{i}$, of dimension {m}. The retractor's scaling is still being absorbed in the extendee's scalar s, as before.

The least slight tweak is that, to ensure that the extendor element gets undone in the reverse order in which it was done, the retractor element has been reversed. This replaces implicit and obscure negation with explicit and clear negation—the negation required to reverse. It establishes the just-displayed *farewell ultimately-enhanced **element**-undoing equation*. Let us use it to discover …

Retraction's commuting properties

Since we are now concerned only with how the retractor element commutes with its retractee, and not with its already-familiar normalization and squaring-up, it shall always be expressed pre-normalized and pre-squared-up.

The most transparent way to do that is to tailor the

orthonormal basis so that the retractee is expressed in just one basis element, as had been done with vector-vector retraction and bivector-bivector retraction. Tho possible, this is not practical for every retraction.

Most practical is to just descend to the standard orthonormal basis. Unfortunately, that is not transparent because the sum of more than two or three basis terms quickly becomes incomprehensible.

Fortunately, for the purpose of discovering retraction's commuting properties, it is possible to be both transparent and practical, like so: simply focus on just one of the retractee's basis terms. Every one of its other terms will have the same dimension (since the retractee is an element); dimension $\{n\}$ let's say, which is what determines commuting properties. So the entire retractee will commute in the same way.

Furthermore, scaling is irrelevant for commuting properties, so we can focus exclusively on unscaled orthonormal basis extensions, like this simple retractee: $(i_1 \triangleleft i_2 \triangleleft i_3 \triangleleft i_4 \triangleleft i_5)$. In general, its retractor will be the scaled sum of multiple basis terms; but, since that retractor is an element too, every one of its terms will have the same dimension ($\{m\}$ we had decided, where $m \leq n$).

Consequently, each of these terms will commute with the retractee in the same way; so we can focus on just one of them, subject to this caveat: It must produce a non-zero result—a zero result would simultaneously commute and neg-commute; so would tell us nothing about commuting properties. The reason is simple: exchanging a retraction's arguments can only change the sign of its result; and $+0 = -0$.

The only way a unit retractor can produce a non-zero result

166

is for every one of its basis vectors to be included within its retractee. Any vector outside would be perpendicular to that retractee, which would sooner or later annihilate that retraction.

So, for maximum simplicity, let us presume this retractor: $(i_4 \triangleleft i_5)$, which is included right inside the tail of our retractee $(i_1 \triangleleft i_2 \triangleleft i_3 \triangleleft i_4 \triangleleft i_5)$. Here is how it becomes manifested in the undoing equation:

$$(s\underline{e}_\perp \triangleleft \underline{i}) \bullet \hookleftarrow \underline{i} \ = \ ((i_1 \triangleleft i_2 \triangleleft i_3) \triangleleft (i_4 \triangleleft i_5)) \bullet (i_5 \triangleleft i_4)$$

The extendor $\underline{i}$ here is $(i_4 \triangleleft i_5)$; the extendee $s\underline{e}_\perp$ is $(i_1 \triangleleft i_2 \triangleleft i_3)$, which is also the result of the undoing. However, it is *not* the result of the retraction, whose sign is determined by the sign required for retractor reversal. Here, that is negative since $(i_4 \triangleleft i_5) = -(i_5 \triangleleft i_4)$; so the result of this retraction is …

$$(i_1 \triangleleft i_2 \triangleleft i_3 \triangleleft i_4 \triangleleft i_5) \bullet (i_4 \triangleleft i_5) \ = \ -(i_1 \triangleleft i_2 \triangleleft i_3)$$

Both sides of the undoing equation reverse the retractor; so the sign of its reversal does not affect commuting properties, which are only concerned with whether the sign *changes* from one side to the other, or not. To say this differently, the retractor's uniform sign of reversal can be absorbed in the irrelevant scalar.

To find out if the sign of commuting changes, just walk the extendor across the retractee, thru its extendee. If that changes sign, then this retraction neg-commutes. If it doesn't, the retraction commutes. Let's try:

$$(i_4 \triangleleft i_5) \triangleleft (i_1 \triangleleft i_2 \triangleleft i_3) \ = \ (i_1 \triangleleft i_2 \triangleleft i_3) \triangleleft (i_4 \triangleleft i_5)$$

There is no sign change here because $(i_4 \triangleleft i_5)$ has even dimensional parity, so induces an even number of negations when walked across any extension. So $(i_1 \triangleleft i_2 \triangleleft i_3 \triangleleft i_4 \triangleleft i_5) \bullet (i_4 \triangleleft i_5)$ commutes, meaning that it equals $(i_4 \triangleleft i_5) \bullet (i_1 \triangleleft i_2 \triangleleft i_3 \triangleleft i_4 \triangleleft i_5)$. The reason is simple: Its undoing expression,

$$((i_1 \triangleleft i_2 \triangleleft i_3) \triangleleft (i_4 \triangleleft i_5)) \bullet (i_5 \triangleleft i_4)$$

… obviously equals its commuted undoing expression,

$$(i_5 \triangleleft i_4) \bullet ((i_4 \triangleleft i_5) \triangleleft (i_1 \triangleleft i_2 \triangleleft i_3))$$

… because the extendor's journey across the retractee, had not caused a sign change. If that journey had caused a sign change, the retraction would have neg-commuted.

This can only happen in a peculiar situation, namely an odd retractor \ extendor element having an odd extendee element to walk across, inducing an odd number of sign changes. Said better, an odd retractor with an even retractee.

So now you know retraction's generic commuting properties: its two arguments usually commute, just like extension's arguments usually do. In particular, any time the retractor is even, its retraction commutes; like extension commutes any time either of its arguments is even. Only when the retractor is odd, and its retractee is even does retraction neg-commute.

Consider how strange that seems: *Extension's* commuting properties depend only on argument dimensions, irrespective of who is extendor, and who is extendee.

Retraction's commuting properties are much more picky. First you must identify who is the retractor. The little guy. Then you must find out if it is *an odd little guy*. Then you must discover if its retractee is even a big guy … I mean *an even big guy*. If so, you know that this retraction neg-commutes. Otherwise it commutes.

Nonetheless, these non-obvious commuting properties derive directly from extension's obvious ones via its undoing. Semantics engenders syntax, you know.

Retraction's expression of projection

(Frank reader advice: I had much fun doing this, and the idea is important, but you may not have much fun watching it. I personally didn't in my final pass because it is unexpectedly intricate for generic elements. To maximize your fun, and get your own idea, your best option may be to rassle with them yourself first, and only scrutinize this section when you need a hint. Otherwise, fly low to *Perpendicularity and Parallelity*, which use the idea and its converse, and shall be important subsequently.)

Geometrically, implicit projection is the very first step for retraction. Algebraically, it generates the scalar separation of the projected retractor, which immediately gets entangled with the separation of its retractee; and then the precise dimensions of the retractee and retractor are promptly lost— only their difference is retained.

Untangling all of that and recovering retractor dimension is the job of explicit projection. Its algebra turns out to be elegant but intricate; and can only be understood after its geometry has been.

Geometrically, projection is the perpendicular shadow of a projector onto its projectee. The spatial expanse of the shadow is determined by the spatial expanse of the projector, of course; but the spatial expanse of the projectee is irrelevant,

169

so long as it is at least large enough to accommodate the shadow. Which it certainly is if the projectee is the retractee, the big element.

If it is *exactly* big enough, projection can resort to an adventitious kludge: use the dimension of the projectee to stand in for the dimension of the projector. This simplifying ploy is what the previous projections—vector-vector specifically, and *n-n* in general—resorted to.

This kind of projection is easy to understand and easy to compute, but it won't work for even such a simple projection as vector onto bivector, v *in B*, which is the only *little-big* projection we humans can visualize. Unfortunately, its geometric simplicity can easily mislead us into ungeneralizable algebra; so we shall need to test the algebra's generality after generating it.

Toward that goal, the first step is to disentangle the separation of the bivector from the separation of the vector's shadow. That is easy by working exclusively with the bivector's direction $I = {}^{\prime}B$, which has unit separation.

Unfortunately, this immediately makes the algebra hard— normalization does not respect summary. For example, it is not true that ${}^{\prime}(3i_1 + 4i_2) = {}^{\prime}(3i_1) + {}^{\prime}(4i_2)$, as would be enlightening to sketch by visualizing the Pythagorean Theorem. (Slightly different direction, slightly different separation.)

So there is no hope of beginning bottom-up, basis-extension by basis-extension, and then transferring their projection to their sum, as had just been done for commuting properties. Instead we must begin top-down, step by step, until the algebra becomes clear.

170

The second step down from the top is to the give the projector's shadow the same dimension the projector has. Here is where we could easily go wrong with vector-bivector projection—its retraction fortuitously produces a result with precisely that property.

So we might initially view its vector result as the projection. However, retraction's squaring-up step reveals that, altho bivector normalization properly gave that vector the shadow's length, it is perpendicular to the shadow.

So the next idea that occurs is that all we have to do is to rotate that vector a quarter turn onto the shadow. That is clearly the final *result* we want for this simple case, but it is not the *action* we need for *m-n* projection in general.

The action we need is to recover projector dimension $\{m\}$ from the dimension of the retraction result, $\{n–m\}$, and the dimension of the projectee, $\{n\}$. Is there any way to do that?

Sure. Relative to the projectee, the retraction result has become a little-element. If we retract that result with the projectee, we obtain dimension $\{n–(n–m)\}$, namely dimension $\{m\}$, as desired. So it looks like, to project a projec*tor*, <u>*tor*</u> let's call it, onto a projec*tee*, <u>*tee*</u>, all we need is this operation:

$$\underline{tor} \text{ } in \text{ } \underline{tee} \text{ } =? \text{ } ({}^{/}\underline{tee} \bullet \underline{tor}) \bullet {}^{/}\underline{tee}$$

That looks suspiciously too simple, so let us test it on the case at hand, namely vector-bivector projection:

$$v \text{ } in \text{ } B \text{ } =? \text{ } ({}^{/}B \bullet v) \bullet {}^{/}B$$

To see if this is even viable, let us try a ridiculously simple example first: ${}^{/}B = I = i_1 \blacktriangleleft i_2$ and $v = 4i_2 + 3i_3$. Without doing any algebra, we immediately know that v's projection into B is the term inside its unit direction, I, namely $4i_2$. The other term, $3i_3$, is perpendicular to I, so it does not contribute to the projection.

Let us see if our new algebra agrees: The innermost retraction, $^/B \bullet v$, becomes $(i_1 \triangleleft i_2) \bullet 4i_2$ plus $(i_1 \triangleleft i_2) \bullet 3i_3$ when multiplied out. The first term becomes $4i_1$; and the second term vanishes.

Consequently, the outermost retraction becomes $4i_1 \bullet (i_1 \triangleleft i_2)$, which equals $4i_2$. Success!—that is what we had concluded. ¿Who would have thought that merely getting the projected *dimension* right would get the *whole projection* right? Here is a picture:

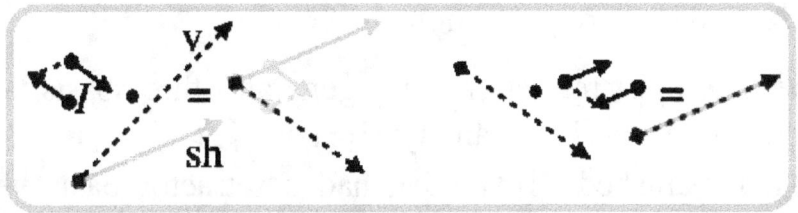

vector-bivector projection

On the left, shown in perspective, is the innermost retraction $^/B \bullet v = I \bullet v$. Vector v here has separation 5; its shadow onto I, sh, shown in gray, has length 4. The retraction produces a vector with that separation, but perpendicular to it, as you see.

On the right is the outermost retraction, namely the previous result retracting of bivector I, which does indeed rotate that vector onto v's perpendicular shadow, as you see again.

Enlightening exercise: Perform these two retractions geometrically: project, normalize, square-up, unextend. I rotated bivector I a quarter-turn on the right, hoping to make that easy. (Hint: this is odd-retractor, even-retractee, so it neg-commutes.) If this was easy, you have nearly mastered retraction. If it wasn't easy, it can become easy with a little more practice.

Let us explore this low-dimensional example with a more realistic projector v, say $3i_1 + 4i_2 + 5i_3$. It has separation $\sqrt{(3^2 + 4^2 + 5^2)}$, a tiny bit over 7.

To facilitate exploration, we shall work with a *pre-normalized* projectee, $I = i_1 \blacktriangleleft i_2$. The advantage is that it puts us back in summary-respecting heaven because each of projection's subsequent retractions respects summary. Consequently, we will be able to analyze each basis projection with I individually, confident that the sum of them represents the entire projection.

The innermost retraction, $I \bullet$ v generates three basis terms, but only two survive, namely $(i_1 \blacktriangleleft i_2) \bullet 3i_1 + (i_1 \blacktriangleleft i_2) \bullet 4i_2$. The term that vanished, $(i_1 \blacktriangleleft i_2) \bullet 5i_3$, had a retractor basis vector outside its retractee, which is precisely how the algebra performs retraction's initial implicit projection.

The terms that survive rotate the effective retractor by a quarter turn to $-3i_2 + 4i_1$ (sketch?), which is precisely how the algebra performs retraction's squaring-up step.

Next, the outermost retraction, $(-3i_2 + 4i_1) \bullet (i_1 \blacktriangleleft i_2)$ rotates the result of the first retraction back to the effective retractor, namely $3i_1 + 4i_2$. This is the final result—the projection of v onto I, with separation $\sqrt{(3^2 + 4^2)} = 5$. There are three important take-aways from this example:

- Basis terms in the projector not included within the projectee are ignored during the innermost retraction. This is how the algebra performs perpendicular

173

projection.

- Terms in the projector that survive are the *implicit* effective retractor. Retraction explicitly rotates them perpendicular to themselves, which is how it squares-up its result against the retractor.

- The outermost retraction merely *un-rotates* that result *reverse-perpendicular* back into an *explicit* effective retractor, which is the result of the projection.

Knowing all of this, a person might think that projection would be more efficiently performed by aborting the innermost retraction right after it has ignored projector basis terms; and just return the projector terms that survive. They constitute the effective retractor, and the actual projection.

That may be more effective, but it is not algebra. It would require a special-purpose algorithm that *explicitly* performs the ignoring and paying-attention that the metric does *implicitly*. Consequently, it might not be faster—an orthonormal metric is blazingly fast.

So let us press on with the algebra, and test it at higher dimensions, say bivector-trivector projection. Again, we will work with a simple pre-normalized unit projectee, $I = i_1 \triangleleft i_2 \triangleleft i_3$, so that we can remain in transparent summary-respecting territory.

In physical space its projector would automatically be inside it, meaning that it would automatically be the effective retractor, and the projection result. So let us presume a much higher space with a four-term bivector projector B, to explore a more realistic scenario:

$$B = 3\, i_1 \triangleleft i_2 + 4\, i_2 \triangleleft i_3 + 5\, i_3 \triangleleft i_4 + 6\, i_4 \triangleleft i_5$$

Inspection reveals one term completely outside the projectee, namely $6(i_4 \triangleleft i_5)$; and one term partially outside, $5(i_3 \triangleleft i_4)$. Both terms will be ignored in the innermost retraction, $I \bullet B$, because they each contain at least one vector perpendicular to I.

Two terms are completely inside I, so will survive: $3(i_1 \triangleleft i_2) + 4(i_2 \triangleleft i_3)$. They constitute the implicit effective retractor and the eventual projection result; but the algebra has no knowledge of that—it just blunders thru the calculation $(i_1 \triangleleft i_2 \triangleleft i_3) \bullet (3\,i_1 \triangleleft i_2) + (i_1 \triangleleft i_2 \triangleleft i_3) \bullet (4\,i_2 \triangleleft i_3)$ and generates $-3i_3 - 4\,i_1$. Please check—this is not going to seem right.

The reason it won't seem right is that the outermost retraction, $(-3i_3 - 4i_1) \bullet (i_1 \triangleleft i_2 \triangleleft i_3)$ generates $-3(i_1 \triangleleft i_2) - 4(i_2 \triangleleft i_3)$, the negative of what we know the retraction result should be. (Please double check.) ¿What went wrong?

What went wrong was the naive projection algebra ($'\underline{tee} \bullet \underline{tor}) \bullet '\underline{tee}$. It worked fine for vector-bivector projection, but gave the wrong sign for bivector-trivector projection.

That is a sure … *sign* … of a neglected reversal that would have been positive in the first situation and negative in the second. Ah ha!—the projector should be reversed:

$$\underline{tor} \text{ in } \underline{tee} \;=?\; ('\underline{tee} \bullet {}^{\curvearrowleft}\underline{tor}) \bullet '\underline{tee}$$

Reversal of a *vector* projector does not change its sign; but reversal of a *bivector* projector does; precisely what is needed to get the projection algebra to work in both scenarios. Wild success!

Nonetheless, I left a question mark in the equation because we still haven't checked it in a more generic scenario. Richard

175

Feynman said to be careful about fooling yourself—"*You are the easiest person to fool*". Before checking, let us pause to consider the take-aways from *this* example:

- *Reversed* basis terms in the projector not included *completely within* the projectee are ignored during the innermost retraction. This is how the algebra performs perpendicular projection.

- Reversed terms in the projector that survive are **not** the implicit effective retractor—their re-reversal is (the *original* effective retractor). Retraction performs *dimension-changing perpendicularity* on the terms, which is how it squares-up its result against the retractor.

- The outermost retraction performs *dimension-recovering reverse-perpendicularity* on that result. This converts it back into an *explicit* effective retractor, which is the result of the projection.

The afore-mentioned *dimension-changing perpendicularity* simplifies to a quarter turn for a unit bivector retracting a vector; and *dimension-recovering reverse-perpendicularity* simplifies to a quarter turn in the opposite direction.

Now to check the generic scenario for *m-n* projection: We can ignore terms the algebra ignores, meaning that we can focus exclusively on projector terms completely within the projectee. These terms will become the individual effective retractors, and their sum will become the projection result.

By again working with a pre-normalized projectee, $i_1 \triangleleft i_2 \triangleleft i_3 \triangleleft i_4 \triangleleft i_5$ let's say, we can focus on just one such term knowing that all other terms behave similarly. Let us try projector term $7(i_1 \triangleleft i_3 \triangleleft i_5)$. All of its vectors lie within the projectee; so we can percolate them against the reversed projector without changing the algebra like this:

$$-(i_1 \triangleleft i_4 \triangleleft (\mathbf{i_1} \triangleleft \mathbf{i_3} \triangleleft \mathbf{i_5})) \bullet {}^{\curvearrowleft}7(i_1 \triangleleft i_3 \triangleleft i_5)) \bullet -(i_1 \triangleleft i_4 \triangleleft (\mathbf{i_1} \triangleleft \mathbf{i_3} \triangleleft \mathbf{i_5}))$$

The percolated term has been emboldened in each unit projectee to emphasize the *algebraic* essence of the projection expression: ***it annihilates the projected element in the innermost retraction; and then recovers it in the outermost retraction.*** Such annihilation–recovery exploits the metric in a useful way for which it was not envisioned. Take a look after the first retraction:

$$7(i_1 \triangleleft i_4) \bullet (i_1 \triangleleft i_4 \triangleleft (\mathbf{i_1} \triangleleft \mathbf{i_3} \triangleleft \mathbf{i_5}))$$

The projected element $(\mathbf{i_1} \triangleleft \mathbf{i_3} \triangleleft \mathbf{i_5})$ has disappeared from the innermost retaction. (And so did the double minus signs of percolation, something that will always happen.) That element is going to get recovered in the final retraction.

In the final retraction … oh … wait … the innermost residue $7(i_1 \triangleleft i_4)$ will need to be reversed first to get the final sign of the projection element right. ¿Why didn't I notice that before? It was because I had accidentally started with projections whose innermost residue was a vector. Their reversal does not change sign.

I was fooling myself. Happily, having checked myself, I now know that the innermost retraction will have to be reversed after it is completed, leaving this final enhanced *m-n* projection expression:

$$\underline{tor} \; in \; \underline{tee} \;\; = \;\; {}^{\curvearrowleft}({}^{\prime}\underline{tee} \bullet {}^{\curvearrowleft}\underline{tor}) \bullet {}^{\prime}\underline{tee}$$

A person might think these two reversals would make the computation cumbersome. That is not so: reversals would be implemented—for computational purposes—not as vector

177

rearrangement, but rather as dimension-dependent negation, which would have negligible effect on computation. For *conceptual* purposes, however, vector rearrangement is essential for understanding reversal.

To sidestep the entire concept of reversal, the aborting algorithm mentioned previously could be used. It returns the projector terms that would survive in the projectee, without proceeding with any metric computations. An optimized algorithm like this might be useful if you need to perform projection often.

Enlightening exercise: Endeavors like this projection effort often begin on a path away from a goal, but manage to arrive at it nonetheless by apparently brilliant cleverness; which is belatedly recognized to have been a brash kludge. The remedy is to never be satisfied:

¿Could the two preliminary reversals be consolidated into a *reverse* percolation and one final reversal? For example …

$$(7(i_1 \triangleleft i_3 \triangleleft i_5) \bullet -((\mathbf{i_5} \triangleleft \mathbf{i_3} \triangleleft \mathbf{i_1}) \triangleleft i_2 \triangleleft i_4)) \bullet -^{\curvearrowleft}((\mathbf{i_5} \triangleleft \mathbf{i_3} \triangleleft \mathbf{i_1}) \triangleleft i_2 \triangleleft i_4)$$

Do you like this better? If so, here is the *truly final, ultimately enhanced m-n projection expression*:

$$\underline{tor}\ in\ \underline{tee}\ =\ (\underline{tor} \bullet {}^{\prime}\underline{tee}) \bullet {}^{\curvearrowleft\prime}\underline{tee}$$

Either one of these *m-n* projections induce the following *m-n* rejec<u>tor</u>-rejec<u>tee</u> rejection:

$$\underline{tor}\ out\ \underline{tee}\ =\ \underline{tor} - \underline{tor}\ in\ \underline{tee}$$

Incidentally, 1-*n* rejection, meaning vector rejection, can be more-directly computed by first extending the vector with the unit *n*-rejectee, which pays attention only to the vector's rejection from that element. To recover that rejection, simply retract that result with the unit rejectee.

For example, consider the bivector and vector considered previously: $'B = I = i_1 \triangleleft i_2$ and $v = 4i_2 + 3i_3$. Without doing any algebra, we immediately know that v's rejection from B is the term *outside* its unit direction, I, namely $3i_3$. The other term, $4i_2$, is parallel to I, so it does not contribute to the rejection.

Consequently, we could sidestep the algebra with an algorithm that simply returns v's basis terms outside its rejectee. But this only works in an orthonormal basis; and in any case the algebra can do it automatically, and maybe even faster.

The algebra pays attention to the rejection like so: $(i_1 \triangleleft i_2) \triangleleft (4i_2 + 3i_3) = 3(i_1 \triangleleft i_2 \triangleleft i_3)$. To extract the rejection from this extension, simply reverse this result and retract it by I: $3(i_3 \triangleleft i_2 \triangleleft i_1) \bullet (i_1 \triangleleft i_2) = 3i_3$. Here is the algebra and a picture of it:

$$v \; out \; \underline{tee} \; = \; {}^\curvearrowright('\underline{tee} \triangleleft v) \bullet '\underline{tee}$$

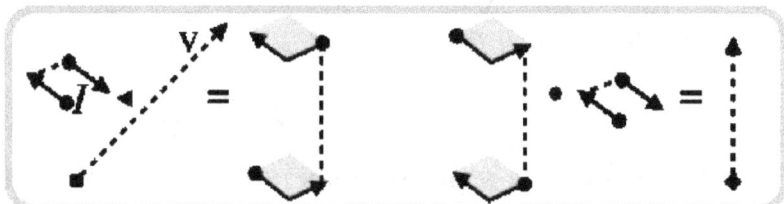

Rejection of a vector from a bivector

On the left you see the extension of I from v paying attention only to v's rejection from I. On the right you see that rejection being extracted via that result *reversed* and retracted by I.

Finger exercise: Speaking of reversal, to exercise your

179

mastery of it show that 1-*n* rejection can also be computed like so:

$$\text{v } out \underline{tee} = (\text{v} \blacktriangleleft {}^{\prime}\underline{tee}) \bullet \,{}^{\curvearrowleft\prime}\underline{tee}$$

Neither of these extension-retraction ploys can be used for other *m-n* rejections because non-vector extensions often return zero even when their arguments are not parallel. In such a case they do indeed have rejection that a zero result would ignore. Which brings us to the final section:

Perpendicularity and parallelity

This is a somewhat intricate topic because there are various degrees of perpendicularity and parallelity. The crucial ones are *complete* perpendicularity and parallelity. If the intricacy of the other kinds begins to bewilder you, skip to the pithy *summary* paragraphs near the end of this section.

To understand *retraction's* expression of perpendicularity and parallelity well, we must first understand *extension's* expression of parallelity. It begins, as with everything else in the algebra, with primitives; and deals intrinsically with *spatial expanse*. This means that it is inherently a *free* concept —the only primitives with such expanse are free vectors. Bound elements inherit perpendicularity and parallelity via that of their free parts.

The parallelity of a free vector with a flat space is traditionally expressed as a scaled sum of the vectors spanning that space. For example, vector v is typically defined to be *parallel to*—or sometimes *within*—the line defined by x if $v = x\text{x}$; it is defined to be parallel to the plane defined by x y if $v = x\text{x} + y\text{y}$; parallel to the space defined by x y z if $v = x\text{x} + y\text{y} + z\text{z}$; and so on.

180

This tedious summary process becomes automated after those spaces become defined by extension: x, x◂y, x◂y◂z. Then their parallelity with v can be checked simply by discovering whether v's extension with them vanishes:

$$v◂x \;=\; xx◂x \;=\; 0;$$
$$v◂(x◂y) \;=\; (xx + yy)◂(x◂y) \;=\;$$
$$xx◂x◂y \;+\; yy◂x◂y \;=\; 0 + 0;$$

and so on.

Said conversely, any scaled sum of the vectors within an extension is a factor of that extension. Consider, for example, the extension a◂b◂c, and the scaled sum $aa + bb$. That sum can replace vector a in the extension, like so: $1/a(aa + bb)◂b◂c = a◂b◂c$. Please check. After you understand how this works, try using that sum to replace b in the extension. Then try replacing c.

¿Did the replacement of c work? Hint: No—such a scaled sum can only replace an extension vector included in the sum. Otherwise the extension would vanish.

The equivalence between a vector being within an element, and its disappearance under extension with that element, provides an expressive way to keep your vectors confined to a subspace of interest.

Suppose, for example, that you are working with a space whose direction is $\underline{i}$. Simply declare, as a nonce axiom, that $x◂\underline{i} = 0$ for all vectors x that you want to keep within $\underline{i}$. (This ploy also works for *points x* you want to keep in a *bound* space $\mathbf{\underline{i}}$: $x◂\mathbf{\underline{i}} = 0$.)

The ploy is not needed to keep vectors inside your global ambient space—the basis itself does that simply by failing to provide any basis vectors beyond your ceiling. Nevertheless, if you wish for analytic purposes to avoid descent to a basis, you may declare formally that every vector extended with your ceiling vanishes.

Encouraged by a vector's parallelity via disappearance under extension, you might hope to extend that ploy to arbitrary free elements. Alas, that idea fails. For example, the extension of u◄v and v◄w vanishes, even tho these bivectors are not parallel whenever u is not parallel to w.

True, they are parallel on one extension factor, namely v, but annihilating parallelity on an extension factor, or even many extension factors, does not make elements parallel. For example, $i_1◄i_2◄i_3◄i_4◄i_5$ and $i_1◄i_2◄i_3◄i_4◄i_6$ are parallel on every extension factor except the last; but they are not considered parallel—not even *partly* parallel—by the vernacular, or by the algebra.

In fact, both the vernacular and the algebra consider these two elements *perpendicular*, as we shall see.

For two free elements to be parallel, they must be *completely* parallel—parallel on *every* extension factor. In general, this must be laboriously checked one vanishing vector after another, if extension is your only tool.

However, there is a peculiar situation in which extension can articulate parallelity all at once; but not as a relation between its two arguments, and not as a relation that vanishes; but rather as a relation that *does not vanish*—a relation between one of its arguments and a result that did not vanish.

For example, if u◄v extended with w◄x did not vanish, it produced u◄v◄w◄x. Both of these arguments are (completely) parallel to that result because *every one* of their extension factors is parallel to it.

182

Unfortunately, this argument-result relation between two extensions is of no help if you do not know about it beforehand. In such a case there is a different tool, retraction —not extension—that can discover if those elements are parallel: simply reject the little-or-same-size element from the big one. If this rejection vanishes, then those elements are certainly parallel—the little element is a pure projection.

Conversely, if the rejection did not vanish, but the projection leading to it did, then those elements are perpendicular. In other words, extension can articulate a clumsy kind of *perpendicularity* via its *rejection* properties; analogous to the way that retraction had just articulated a clumsy kind of *parallelity* via its *projection* properties.

However, retraction is far more expressive about *perpendicularity*, which it expresses complementary to the way extension had expressed parallelity. That journey begins with a vector *retraction* that vanishes, rather than an extension that does.

Suppose, for example, that $v \bullet (x \triangleleft y)$ had vanished. Geometrically, the reason is that v had no preliminary projection onto $(x \triangleleft y)$; v was *equiangular* to that bivector— perpendicular on *every vector* within it.

Such total perpendicularity is corroborated by the algebra: $v \bullet (x \triangleleft y) = v \bullet x\ y - v \bullet y\ x$. Since x and y are independent, the only way this sum can vanish is for every term in it to vanish, meaning that $v \bullet x = 0$, and $v \bullet y = 0$. Whence v is perpendicular to x, and perpendicular to y; and of course perpendicular to every scaled sum of them.

Strengthening exercise: Repeat this analysis on $v \bullet (w \triangleleft x \triangleleft y \triangleleft z) = v \bullet w(x \triangleleft y \triangleleft z) - v \bullet x(w \triangleleft y \triangleleft z) + v \bullet y(w \triangleleft x \triangleleft z) -$

v•z(w◄x◄y) ¿Are the terms in this alternating sum independent? If so, must *every* term vanish to make this retraction vanish? What does that mean geometrically?

Encouraged by a vector's *perpendicularity* via disappearance under *retraction*, you might have hoped to extend that ploy to arbitrary free elements. But having just seen the failure of that idea for *parallelity* under *extension,* you may have been discouraged.

Happily, retraction provides reason for encouragement: the idea succeeds for a *retractor* element that causes its result to disappear — that retractor is indeed perpendicular to its retractee.

This is a little strange because a retractor needs *just one* of its extension vectors to be perpendicular to its retractee for the entire retraction to vanish. This complements the way that an *extendor* needs just one extension vector *parallel* to its extendee for the entire *extension* to vanish.

(If *just-one* is not obvious for retraction, like it is for extension, simply recognize that generic retraction is iterated retraction by the extension vectors in a retractor. Neg-commute the perpendicular vector to the innermost iteration. It immediately annihilates that retraction.)

Consequently, an element need not be *completely perpendicular* to be perpendicular; unlike the way that an element needs to be *completely parallel* to be parallel.

This may be easiest to understand for two planes in physical space, say $i_1 ◄ i_2$ and $i_2 ◄ i_3$. They are parallel on i_2, but no one would ever call them parallel. They are perpendicular on i_1 and i_3, and everyone would call them perpendicular.

And so would the geometry and the algebra. The geometry would say they have no planar projection onto each other so they are perpendicular—their retraction aborts before it gets started.

True, they have *line* projection, but the algebra would say that doesn't count—the undoing equation requires the projected effective retractor to have the same dimension as the unprojected retractor. Otherwise dimension would not be well defined.

(Incidentally, all of this explains why *rejection* may be implemented as subtraction of *projection*, even tho the projection may have only been perpendicular in just one direction. Ponder perpendicular planes.)

Before leaving the retractor-retractee perpendicularity of disappearing retraction, please note that the perpendicularity of a vector in a *retractee* to its retractor says nothing whatsoever about retractor-retractee perpendicularity.

Indeed, a retractee absolutely must contain vectors perpendicular to its retractor to produce a result—otherwise it could not engage the undoing equation: the extendee \ retraction-result, $s\underline{e}_\perp$, in there is completely perpendicular to the retractor.

Its perpendicularity constitutes retraction's disambiguation —its *squaring-up*—the step after projection and normalization. Which means that we are now advancing beyond a projection that failed, to one that succeeded and has something to square-up against—we have moved beyond disappearing retraction.

We are still in perpendicular territory, not a retractor-retractee kind, but rather a retractor-*result* kind. And that is somewhat of a puzzle because the non-vanishing result of $v \bullet (w \triangleleft x \triangleleft y \triangleleft z)$, for example, is $v \bullet w(x \triangleleft y \triangleleft z) - v \bullet x(w \triangleleft y \triangleleft z) + v \bullet y(w \triangleleft x \triangleleft z) - v \bullet z(w \triangleleft x \triangleleft y)$, which fails to inform as transparently as its vanishing had.

Geometrically, we know that this four-term alternating sum —if it hasn't vanished—*must* be completely perpendicular to v because the undoing equation had told us: a retraction result *must* be equiangular to its retractor vector in order to be well defined.

Algebraically, however, the non-disappearance of this sum, as written, gives no clue about its perpendicularity—every term in it has a part parallel to v and a part perpendicular. It would seem almost magic for the parallel parts to all disappear.

We already derived that magic; and we know that this sum can be more transparently written as $v \bullet w(x_\perp \triangleleft y_\perp \triangleleft z_\perp) - v \bullet x(w_\perp \triangleleft y_\perp \triangleleft z_\perp) + v \bullet y(w_\perp \triangleleft x_\perp \triangleleft z_\perp) - v \bullet z(w_\perp \triangleleft x_\perp \triangleleft y_\perp)$, meaning that it is *completely perpendicular* to v—*every* vector in it is perpendicular.

Which makes the more general perpendicularity of non-vector, non-vanishing retraction easy to understand. To begin, let us revisit an example already seen, namely $(v \triangleleft w) \bullet (x \triangleleft y \triangleleft z)$:

"We know that the first iteration had produced a bivector perpendicular to w—*every vector* in that plane is perpendicular to w—and then, *within that plane*, the second iteration had produced a vector perpendicular to v. Consequently that final vector is perpendicular to both v and w, to $v \triangleleft w$." In general a retractor element is always completely perpendicular to its retraction result, and vice

186

versa.

That is the complete story on perpendicularity and parallelity of retraction and extension. Here is a summary:

Vanishing vector extension informs about vector-extendee parallelity. Other vanishing extensions do not inform about parallelity because it must be *complete*, which requires laboriously checking vanishing of *every* vector in an element, one by one. Non-vanishing extensions can check (*complete*) parallelity all at once; but only for the limited case of extendor \ extendee-result.

Conversely, vanishing vector *retraction* informs about vector-retractee *perpendicularity*. Other vanishing retractions inform similarly about retractor-retractee perpendicularity; but this kind of perpendicularity is incomplete—it requires just one retractor vector perpendicular to its retractee. Non-vanishing retractions can check *complete* perpendicularity— perpendicularity on *every* vector—but only for the limited case of retractor-result.

Extension can articulate a clumsy kind of perpendicularity via a rejection whose projection had vanished. Retraction can articulate a clumsy kind of parallelity via a projection whose rejection had vanished.

Don't worry if you did not understand many of the intricacies in this chapter—I personally have to occasionally come back and check them myself.

The important take-aways are that bound retraction does not work, except for retraction by a point in a confining space, which produces the free part. A vector retracted with anything perpendicular to itself vanishes. A retractor is completely perpendicular to its retraction result, which is completely parallel to its retractee. Orthonormal vector retraction is trivial when juxtaposed: $(i_a...i_b) \bullet i_b = i_a...$ and $i_a \bullet (i_a...i_b) = ...i_b$. And finally, retraction of two generic elements is iterated vector retraction by the little element.

If you got that much out of this chapter, you got the essence. The sailing will be smoother in the next chapter, but will enter high-dimensioned intricacies of little immediate relevance to physical space. The sailing in the *Synthesis* chapter will be fairly smooth thruout, which is fortunate because that is the crucial chapter.

Adding and Composing

Here is a conundrum: Expression of anything in a basis is *inherently composite*—a sum of distinct things. And yet many of the things thus expressed coalesce to something *intrinsically non-composite*—**singular** let's say—a bound vector, for a familiar example.

Distilling singularity

Let's rassle with that example. A bound vector can be expressed in an *o p q* basis with the following *basis composition*, specifically, a **2**-*composition* because each term is an extension of **2** bound primitives.:

$$o(o \cdot p) \;+\; p(p \cdot q) \;+\; q(q \cdot o)$$

The *Extending* chapter had already explained that a sum of bound vectors like this might not coalesce to a single one. This happens in physical space when the vectors do not intersect, not even "at infinity". However, in an *o p q* plane they always do; so we would naturally expect this sum to always coalesce to a single bound vector.

Checking for such singularity always proceeds in the same way: begin coalescing the basis sum one term at a time until you cannot coalesce any more. If you haven't exhausted the composition, you know its sum is *plural*. To discover just how plural, continue from where you left off, trying to

189

coalesce until you cannot. And so on.

For the case at hand, the first two terms, $o(o◄p) + p(p◄q)$, have a common extension factor, p; so they promptly coalesce to $(oo–pq)◄p$. That was almost too easy; but coalescing what remains, $(oo–pq)◄p + q(q◄o)$, may not look so easy.

True, the task is a little intricate, but it is actually straightforward once you recall from the last chapter that *a scaled sum of extension factors* is itself a factor of that extension. So $(oo–pq)$ is a factor of $q(q◄o)$.

If I perform this for you, it will seem mysterious and difficult. If you perform it for yourself, it will seem obvious and easy. *Obvious-and-easy* coalescing is crucial for everything in this chapter; so here is an …

Edifying exercise: Show that $q(q◄o) = –q/p\ (oo–pq)◄o$. Please rassle this down until you understand it completely—it is easier than it looks owing to annihilating self-extension. The concept is what is important—we will seldom descend to such detail hereafter. Now, however, that detail makes the basis composition finally coalesce:

$$(oo–pq)◄p\ +\ (oo–pq)◄{–q/p}\ o\ =\ (oo–pq)◄(p – q/p\ o)$$

Success! This is a single bound vector, dimension $\{\mathbf{2}\}$, because the final result on the right is an extension of two points. We know that because the sum of points is always a point …

… oh … wait … uhmn, if o had equaled p, the left point sum $(oo–pq)$ would have actually been plural, a free vector. But that is okay because it is extended with a single point $(p – q/p\ o)$, which still produces a single bound vector. Oh … well, if q/p had been 1, that would have been a free vector too.

190

In that case the basis composition would actually represent something *plural*—*two* separate-but-opposite bound vectors, a free bivector, dimension {**2**, **2**}. This happens when $o=p=q$. In this case the extension has zero magnitude because both of its primitive factors did (with no appeal to a metric). You already saw this scenario in the *Extending* chapter for the case where o p q were all 1.

Refreshing exercise: Recall that was a head-on-tail cycle of three bound vectors that could be shown to be separate-but-opposite sums in three ways—one way for each vertex. Re-sketch all three scenarios, and then show that the enhancement to a non-unit scalar still launches this free bivector thru the vertices, but with rescaled sides—rescaled area-separation.

The preceding distillation just revealed a truly serious conundrum, rather than a mere paradox: To distill singularity, we not only must coalesce extensions as much as possible; but also, after coalescing, we must check for the singularity or plurality of the primitive sums within them.

This is crucial because *primitive sums are the only sums that remain in the factors of an element*, by upcoming definition. One would hope that they would be dependably singular so they could be used as the foundation for all singularity. Happily, the move to a free-as-possible basis does just that—it makes the sum of all primitives *syntactically* singular.

So, this book shall hereafter represent all numbers using a free-as-possible bound basis, an orthonormal one. Such a basis for physical space would have primitives o i_1 i_2 i_3. They generate an extended basis containing $2^4 = 16$ elements, as the *Extending* chapter explained. We will start there, and then move up to higher spaces to enhance our understanding.

Even tho free vectors i_1, i_2, i_3 have become *syntactically* singular by installation in the basis, they must remain *semantically* plural in order for the full algebra to make the crucial free-versus-bound distinction. Doing so requires a more nuanced distinction than *singular versus plural*, namely *elemental* versus **mixed**.

Primitive sums are elemental

Elemental means syntactically singular but perhaps semantically plural, or not. The most syntactically singular thing in our algebra is the origin *o*. It can be translated and weighted wherever a point is needed, neither of which has any effect on its singularity.

To reinforce that, it is made *semantically* singular by being assigned a singular dimension of {**1**}. This is a *semantic axiom* that defines a property, rather than a *syntactic axiom* that defines a grammatical relation.

Next, free vectors were made syntactically singular by being installed as single elements in the primitive basis. However, to preserve their meaning they must be made *semantically* plural by being assigned a plural dimension of {**1**, **1**}-*without magnitude*. This is abbreviated to {*1*} for convenience, and also for conformity to the conventional algebra, and finally in recognition of the cohesive nature of *without-magnitude* plurality.

All primitives have now become syntactically singular *elements*; regardless of whether they are semantically singular, with dimension {**1**}; or semantically plural, with dimension {*1*}. Then, extensions of elements are also elements; and may be semantically singular, with dimension $\{n\}$; or semantically plural, with dimension $\{n\}$, meaning dimension $\{\mathbf{n}, \mathbf{n}\}$-*without magnitude*.

Finally, *mixed numbers* are sums of elements that fail to coalesce to a single element, indicated by having a formal *mixed dimension*. Defining that dimension well is the main job of this chapter.

That job would be much facilitated by knowing that a sum of primitives is always as *elemental* as the primitives themselves—always *syntactically singular*. This does not happen in a pure point basis, as just seen; so it is prudent to check elementality for the three different kinds of free-as-possible primitive sums:

First, the sum of any two free vectors, u+v, is a single free vector w. Each of these vectors is distinguished in the primitive basis by an absent origin: $0o + s_1i_1 + s_2i_2 + s_3i_3$. The dimension of such a sum is always $\{1\}$, regardless of the number of free vectors within it.

What makes such syntactic *singularity* obvious, ironically, is the semantic *plurality* of free vectors: when the head of one is placed on the tail of the other, they cancel, leaving a head subtracted from a tail, another free vector—another *primitive element*. This process continues until you run out of free vectors to add.

Next, the sum of any two weighted points, $aa+bb$ is a single weighted point cc, provided that weight $a+b$ had not vanished: $ao+a + bo+b = co+c$, where scalar $c = a+b$ and vector c = a+b. This constitutes the most general primitive basis composition—the most general kind of primitive element: $oo + s_1i_1 + s_2i_2 + s_3i_3$, for non-zero o. The dimension of such a sum is always $\{1\}$, regardless of the number of free vectors within it.

And again, what makes this *singularity* obvious is the semantic *plurality* of free vectors: when the tail of the first basis vector is placed on the origin, that tail cancels the origin, leaving a head. When the tail of the next free vector is placed on that head, they cancel. And so on, finally leaving a point — another *primitive element*.

If weight $a+b$ had been 0, the origin would have vanished, leaving free vector c = a+b, another primitive element. This is the strange way that syntactic singularity accurately induces semantic plurality — not by packaging things into an intrinsically composite bundle, as needed in a point basis; but rather by just discarding something with singular dimension $\{1\}$, leaving plural dimension $\{1\}$ as residue.

If that does not seem strange to you, then you haven't quite understood free vectors, to paraphrase what Bohr said about quantum mechanics.

Finally, the sum of a weighted point $ao+a$ and a free vector b is another weighted point $co+c$, another primitive element, where scalar $c = a$ and vector c = a+b. All of this syntactic singularity provides the elemental primitive foundation needed to begin …

Distilling elements

Adding mixed numbers $\underline{M}$ and $\underline{N}$ using an extended basis is absolutely trivial: each number is represented as a scaled sum of various basis elements, of various dimensions, bound or free. Their sum is effectively a *multiple pure scalar sum* — the scalar coefficients of corresponding basis elements are added together. That's it.

This is the venerable four-century-old Cartesian ploy generalized from scalar lengths to scalar weights, scalar length-separations, scalar areas, scalar area-separations, scalar volumes, scalar volume-separations … and so on.

> (By the way, this exposes the dimensional obfuscation that mathematical convention has boxed itself into with so-called "R^n". ¿R^n of what? Weights? Lenths? Length-separations? Areas? Area-separations? … on and on.
>
> This is always the problem with **We-define** distinctions, dimension in this case. Semantically engendered **algebra-derived** distinctions always inform better. Grassmann's extension got us started with dimension, but we have been awfully slow about expressing it more informatively; partly because he himself failed to make the semantically engendered free-versus-bound dimensional distinction—he conflated both into their numeric "*order*".
>
> Subsequent authors preserved that fuzziness under the new name of *grade* owing to raw intellectual inertia. Clifford did not suffer from that—he was a gentle kind of iconoclast—and had he lived I believe he would have eventually formalized that distinction. ("If he had lived, we might have known something" is displayed on the title page to his *Mathematical Papers*, an accolade Newton had launched about Roger Cotes.))

The multiple pure scalar sum ignores dimension, and ignores freedom and bondage, but gets all of that right nonetheless via the massive parallelism of matching basis elements.

We just saw an example of such parallelism in action: the parallel addition of primitives—the basis origin and free vectors—enables the basis to get the freedom and bondage of those sums right without having to explicitly bundle-up free vectors.

So, if addition were the only operation in the algebra, there would be no need to notice dimension, notice freedom or bondage, notice the individual elements represented by a basis sum. Simple matching of basis elements would make those distinctions implicitly. However, other operations —*particularly computation of intersection*—do need to notice those distinctions explicitly; and of course *We* do too —*especially We do*—in order to use the algebra to represent reality.

Noticing elements

Individual elements are the hardest things to notice because they require noticing everything else. First, dimension must be noticed because only terms having the same numeric dimension (same *order*) can coalesce with each other. This means that basis n-compositions of distinct n must be distilled distinctly,

... which immediately disallows an operation all newcomers to geometric algebra attempt—trying to coalesce two vectors, a free one, v say; and a bound one, **w**, by adding them: v+**w**. Coalescence is not possible because their sum has irreducible mixed dimension $\{1, 2\}$. You and I would never try this now, sophisticates that we have become; tho I did try as a newcomer so long ago.

Because noticing dimension is so crucial, we need operations that do so, making all possible dimensional distinctions:

- $^n\underline{M}$ the free n-composition of $\underline{M}$.
- $^{\mathbf{n}}\underline{M}$ the bound **n**-composition of $\underline{M}$.
- $^n\underline{M}$ the free and bound n-composition of $\underline{M}$.

196

The first two operations notice, as a crucial part of their dimensional distinctions, *italic* freedom *n* and **bold** bondage **n**. However, in a free-as-possible basis that is trickier than it may seem because a bound element almost always requires a free one to translate it from the origin to its confining space.

Unfortunately, only a free element *parallel* to the bound one can do that. So if it is not parallel, it must be decomposed into a part that is parallel and a part that is perpendicular. The parallel part gets entrenched with the bound terms in the composition. The residual free perpendicular part becomes independent.

Which raises this question: ¿Do the dimension-extracting operations $^n\underline{M}$, $^n\underline{M}$, $^n\underline{M}$ return the basis terms before or after the parallel-perpendicular, entrenched-residual decomposition?

The answer is *before* because they are low-level operations specifically intended to enable that decomposition. After it has been performed there are two higher-level operations that return the residual free terms, and the translated bound terms:

- $^F\underline{M}$ residual perpendicular free elements in $\underline{M}$.
- $^B\underline{M}$ bound elements in $\underline{M}$ after parallel translation.

To specifically discover an n-dimensioned residual free element, or a translated bound elements in $\underline{M}$, perform $^n\underline{M}$ first.

To make all of this clear, let us begin distilling basis compositions one dimension at a time until we completely understand how that works. Basis 1-compositions have already been understood—they all coalesce to either a single point, dimension $\{\mathbf{1}\}$, or a single free vector, dimension $\{1\}$, nothing else. So we proceed with …

Distilling basis 2-compositions

Let us start with a transparent example and successively enhance it until we understand how to distill all 2-compositions in physical space; and then proceed onward to arbitrary spaces. After that we can repeat the process for 3-compositions, and on up.

Here is a simple but non-trivial 2-composition: $i_1 \triangleleft o + i_1 \triangleleft i_2$, an origin-bound vector plus a free bivector parallel to it (parallel on i_1). That parallelity allows this sum to coalesce to a single bound vector, $i_1 \triangleleft (o+i_2)$, dimension $\{2\}$. This vector points in the direction of i_1, and is bound from the headpoint of i_2, namely $i_2 = o+i_2$.

That was easy to understand, so let us enhance it slightly by adding $i_2 \triangleleft i_3$. This free bivector is perpendicular to the bound vector, so it cannot further translate that vector in any way. Indeed, it is a distinct element algebraically because it has no common extension factor with the bound vector, unlike the first bivector. Here is a picture:

$$(i_1 \triangleleft o + i_1 \triangleleft i_2 = i_1 \triangleleft(o+i_2)) + i_2 \triangleleft i_3$$

Simple 2-composition with 3 terms.

On the left, packaged together, are the first two terms, a vector bound thru the origin plus a free bivector parallel to it. They coalesce to a vector bound thru the headpoint of i_2; a merging that is geometrically obvious, like so: the left end of the bivector annihilates the origin-bound vector, leaving the other end as residue, a simple translation. You have seen this before.

It is then added, on the right, to a free bivector perpendicular to it, leaving two terms: a bound vector *thrust* plus a free bivector *twist*, dimension {**2, 2**}. This is a geometric *minimal form*, already explored in the *Extending* chapter.

Crucial minimal enlightenment: This 2-composition, $i_1 \triangleleft o + i_1 \triangleleft i_2 + i_2 \triangleleft i_3$, can also coalesce into a different *algebraic* minimal form, namely $i_1 \triangleleft o + i_2 \triangleleft (i_3 - i_1)$. This is still the sum of a bound vector plus a free bivector, dimension {**2, 2**}; and it still equals the previous sum. Try sketching it. ¿Why isn't it considered the *geometric* minimal form?

It isn't so-considered because the free term that translates the bound term has not been entrenched within it. Which brings up an obvious distilling principle, in retrospect:

Entrench a free translator with its bound translatee.

Such disciplined assimilation, as will become clear, helps to cleanly separate what is intrinsically bound from what is intrinsically free in a basis composition. In doing so, it also helps to keep basis compositions as elemental as possible.

The foundation for this idea began with primitives: the free vectors that translate the origin in a primitive basis composition are effectively entrenched with it, making it elemental—such a composition has singular dimension {**1**}, not plural dimension {**1**, 1}, and *especially* not dimension {**1**, 1, 1, 1}.

With this principle in mind, let us further enhance our 2-

composition by adding $i_2 \triangleleft o$, thereby generating number $\underline{N} = i_1 \triangleleft o + i_2 \triangleleft o + i_1 \triangleleft i_2 + i_2 \triangleleft i_3$. Entrenching the translators within their translatees leaves two bound vectors, $i_1 \triangleleft (o + i_2) + i_2 \triangleleft (o + i_3)$. Here is a picture:

$$i_1 \triangleleft (o+i_2) \;+\; (\; i_2 \triangleleft o \;+\; i_2 \triangleleft i_3 \;=\; i_2 \triangleleft (o+i_3) \;)$$

Simple 2-composition with 4 terms.

On the left you see the bound vector previously generated. On the right, packaged together, you see another bound translatee plus its free translator. They coalesce to a translated bound vector in the same way the previous translatee + translator had, like so: the bottom end of the free bivector annihilates the origin-bound vector, leaving its top end as residue.

There is good news and bad news about this figure. The good news is that the translators were entrenched with their translatees. The bad news is that this was done prematurely, leaving bound terms not minimal, and free terms excessively minimal (absent in fact). In other words, intrinsically bound had not been cleanly separated from intrinsically free. Which brings up a second, not-so-obvious distilling principle:

Coalesce bound translatees before translation.

For the case at hand, this requires the two bound terms, $i_1 \triangleleft o + i_2 \triangleleft o$, to coalesce to $(i_1+i_2) \triangleleft o$ before translation. This would be a vector bound from the origin, pointing straight right in the coordinate frame, halfway between i_1 and i_2, with length $\sqrt{2}$ (the separation of its free part).

200

The free terms parallel to such a bound vector would translate it to its confining space. The terms perpendicular would be the intrinsically free residue of this composition. The problem is that they are entangled here, as they nearly always are.

It is fair to say that disentangling perpendicular from parallel is the crux of basis distillation in the full geometric algebra.

For the case at hand, such disentangling is geometrically obvious: the free parallel part of the composition must translate the coalesced bound vector halfway between the two previous ones—its equivalent sum—when they are slid so their tails are as close together as possible. Let us work carefully thru that geometry to anticipate what the algebra must generate:

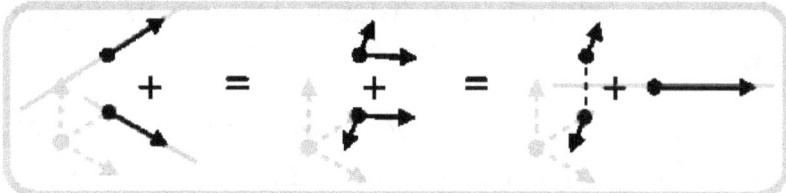

Disentangling free from bound geometrically.

On the left are the two bound vectors that had been mistakenly generated by premature translation. The top one has been slid along its confining line so that its tail is as close as possible to the other tail.

In the middle these bound vectors have been decomposed into parallel parts pointing right; and anti-parallel parts perpendicular to them; as explained in the *Extending* chapter.

On the right is the final sum, a bound-vector thrust plus a free-bivector twist. The thrust is simply $(i_1+i_2)\triangleleft o$ translated by the part of the free composition parallel to it. The twist is the free residue perpendicular to it.

This is all geometrically obvious; but the peculiar relation of the thrust and twist to this coordinate frame, as you see, makes the algebra non-obvious and intricate. The next job is to make that algebra obvious; tho it will remain intricate, of course, in order to properly inform.

Extracting compositional parallelity

Having coalesced the bound basis composition to a free vector extended from the origin, $\mathbf{v} = v \triangleleft o$, the next job is to extract its translator—the free expression parallel to it.

To do so, the projection developed in the previous chapter provides the wrong kind of parallelity. That was *little-big* parallelity—*inside* parallelity—denoted by the *in* operator. The parallelity we need now is the *big-little* kind—*including* parallelity—which shall be denoted by the *include* operator. Take a look at the difference:

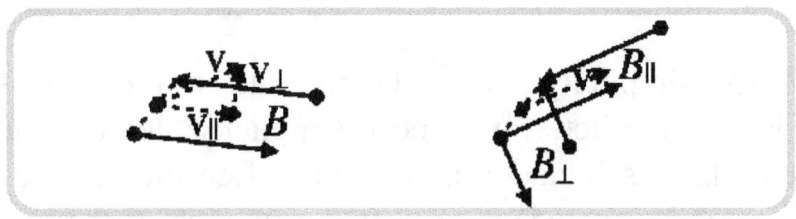

in-out decomposition vs include-exclude decomposition

On the left in this figure you see a vector v, representing the free part of the coalesced bound terms in the composition; and a free bivector B representing the coalesced free terms. (Free 2-terms always coalesce to a single bivector for physical space; but may not for higher spaces, as we shall see shortly).

We need to find the part of B parallel to bound $\mathbf{v}$—meaning parallel to free v—but the projection of v into *B* on the left is of no use for that. What we need is *B including* v, denoted on the right as $B_{\parallel}$. This becomes the translator of bound **v**.

(An excellent tactic at this point is to rassle with this yourself for awhile, which should clarify the following exposition. If it becomes unclear, pausing to rassle should make it clear.)

The residue left after translating **v** is *B excluding* v, denoted as $B_{\perp}$, perpendicular to bound **v** and its inclusion. It is what $^F\underline{N}$ would return, called the *twist* for a 2-composition.

B's exclusion and inclusion of v sum to *B*: $B_{\perp} + B_{\parallel} = B$; just as v's projection and rejection of *B* sum to v: $v_{\perp} + v_{\parallel} = v$.

Inclusion is size-wise complementary to projection, an idea that begins this way: Whereas projection pays no attention to the separation of its higher-or-same-dimensioned projectee; inclusion pays no attention to the separation of its lower-or-same-dimensioned includee. This means that the includee must first be normalized to a unit, just as the projectee had been.

So, for simplicity let us begin with a pre-normalized includee, i. In fact, for utmost simplicity, let us initially suppose that its includor is a generic free element already parallel to i, denoted as $\underline{e}_{\parallel}$. In such a case, the *include* operator takes a trivial form:

$$\underline{e}_{\parallel} \; include \; i \;\; = \;\; \underline{e}_{\parallel}$$

In other words, if i is already included in $\underline{e}$, the *include* operator simply returns $\underline{e}$. This may not seem an advance, but it actually is because in such a case $\underline{e}_{\parallel} = \underline{e}_{\perp} \triangleleft i$, where $\underline{e}_{\perp}$ is a free element perpendicular to i, of decremented dimension from $\underline{e}_\parallel$.

We have already seen $\underline{e}_\perp {\blacktriangleleft} i$ many times; and we know that if we retract it by i, we recover $\underline{e}_\perp$. But we need to recover $\underline{e}_\parallel$ instead, for the *include* operator. That is easy by extending from i our recovery of $\underline{e}_\perp$, namely $(\underline{e}_\parallel \bullet i)$. We have just stumbled onto the important ***extension redoing equation***:

$$(\underline{e}_\parallel \bullet i) {\blacktriangleleft} i \ = \ \underline{e}_\parallel \ = \ i {\blacktriangleleft} (i \bullet \underline{e}_\parallel)$$

This simply says that undoing extension of a unit vector residing inside an element can be redone by extending by that vector; an idea I would hope might be geometrically obvious by now. The idea can be generalized to a generic unit element $\underline{i}$ of dimension less-or-equal to the dimension of $\underline{e}_\parallel$ like so:

$$(\underline{e}_\parallel \bullet \underline{i}) {\blacktriangleleft}^{\curvearrowleft} \underline{i} \ = \ \underline{e}_\parallel \ = \ {}^{\curvearrowleft}\underline{i} {\blacktriangleleft} (\underline{i} \bullet \underline{e}_\parallel)$$

Reversal is needed to ensure that redoing is performed in the reverse order to undoing. This generic redoing equation is exactly complementary to the generic undoing equation, like so: perpendicular element $\underline{e}_\perp$ has been replaced by parallel element $\underline{e}_\parallel$; and extension and retraction have been swapped.

Such redoing performs the *include* operation for the boundary case of a unit-element inclu*dee* already included in a generic inclu*dor*. That encourages us to think that it might work in general, like this:

$$\underline{dor}\ include\ \underline{dee} \ = \ (\underline{dor} \bullet {}^{!}\underline{dee}) {\blacktriangleleft}^{\curvearrowleft !}\underline{dee}$$

… where the numeric dimension of the includor is at least as big as that of its includee. This does indeed work, and the reason is clear:

The includor's initial retraction by a unit-includee reduces the separation of its lower-dimensioned result by the cosine of

204

the angle between them. Then, extension re-creates the includor at that reduced separation in the direction of the includee.

Such reducing re-creation should become clear by carefully working thru the details using the previous figure as an example:

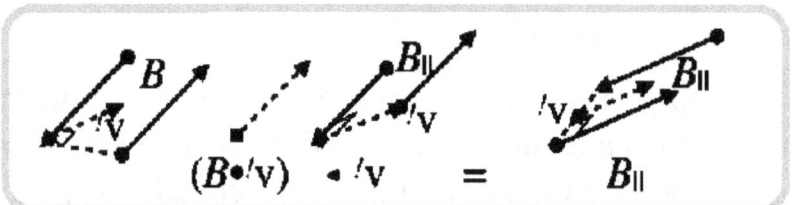

Include: a reducing retraction, a re-creating extension.

The includor here, shown in perspective, is bivector B on the left. The includee is vector v, shown normalized. All bivectors have been squared up, and their sides given unit distance apart; so the length of their sides represents their separation.

That separation, reduced by the cosine of the angle between v and B, is acquired by the result of the preliminary retraction, a free vector parallel to B and perpendicular to v.

When $'$v subsequently extends that vector, it recreates B in the direction of v, at that reduced separation. This is B's *inclusion* of v, denoted as $B_{\parallel}$. It is shown again on the right, rotated a quarter turn in its plane; and displayed with v inside it, as had been done in the previous figure.

To watch the new *include* algebra in action, let us deploy it on the current number $\underline{N} = i_1 \triangleleft o + i_2 \triangleleft o + i_1 \triangleleft i_2 + i_2 \triangleleft i_3$.

First we need the free part of the bound vector. The obvious way to get it is to first extract the bound terms and then take their free part: $^{[]2}\underline{N}$. However extraction of bound terms is not needed because the free-part operator ignores free terms, so all we need is $^{[]}\underline{N} = i_1 + i_2$. Next we need to get the free terms: $^2\underline{N} = i_1 \triangleleft i_2 + i_2 \triangleleft i_3$. Now we can deploy the include operator:

$$^2\underline{N}\ include\ ^{[]}\underline{N}\ =\ ((i_1 \triangleleft i_2 + i_2 \triangleleft i_3) \bullet {}'(i_1+i_2)) \triangleleft \overset{\hookleftarrow}{}{}'(i_1+i_2)$$

The normalized free part of the bound terms, $'(i_1+i_2)$, is $1/\sqrt{2}$ times that free part. This scalar appears twice; so the final retraction-extension will be multiplied by $1/2$. We can absorb the sign of reversal in that scalar, which in this case does nothing. It is convenient to keep that scalar out of the way until the end.

Intricate algebra exercise: Do this retraction-extension. You should get $i_1 \triangleleft i_2 + (i_2 \triangleleft i_3)/2 + (i_1 \triangleleft i_3)/2$, which translates $(i_1+i_2) \triangleleft o$ when added to it. Such addition coalesces *much* easier if you retreat to before the extension, right after the retraction: $(i_1-i_2-i_3)/2 \triangleleft (i_1+i_2)$. (Notice the duplicate extension factor (i_1+i_2)). Now when you do the addition you get $(i_1+i_2) \triangleleft (o+(-i_1+i_2+i_3)/2)$.

This is indeed the bound vector previously derived geometrically, but launched somewhat leftward of where it was shown. It is what $^B\underline{N}$ would return, called the *thrust* for this 2-composition.

Extracting compositional perpendicularity

After having extracted the free terms parallel to the bound ones, the terms perpendicular to them are easy to extract—just subtract: $^2\underline{N} - (^2\underline{N}\ include\ ^{[]}\underline{N}) = (i_1 \triangleleft i_2 + i_2 \triangleleft i_3) - (\ i_1 \triangleleft i_2 +$

$(i_2 \blacktriangleleft i_3)/2 + (i_1 \blacktriangleleft i_3)/2$.

This equals $(i_2-i_1)/2 \blacktriangleleft i_3$, which is indeed perpendicular to (i_1+i_2), as you may check by a vanishing retraction (hint: expand this extension first). This is the twist that had previously been derived geometrically.

Generalizing this tactic to a generic exclu*dor* excluding an exclu*dee* of less-or-equal dimension gives ...

$$\underline{dor} \text{ exclude } \underline{dee} \ = \ \underline{dor} - (\underline{dor} \text{ include } \underline{dee})$$

This is the most convenient way of finding the exclusion if you have already generated the inclusion. However, there is a seductively more elegant and illuminating way that does not require the inclusion first.

It begins in a way complementary to how inclusion had begun, like so: When an element, denoted as $\underline{e}_\perp$, is already perpendicular to a smaller unit $\underline{i}$, the *exclude* operator takes a trivial form:

$$\underline{e}_\perp \text{ exclude } \underline{i} \ = \ \underline{e}_\perp$$

Again, this may not seem an advance, but it actually is because the algebra for such an operation has already been developed and extensively tested—it is simply the generic undoing equation:

$$(\underline{e}_\perp \blacktriangleleft \underline{i}) \bullet {}^{\curvearrowright}\underline{i} \ = \ \underline{e}_\perp \ = \ {}^{\curvearrowleft}\underline{i} \bullet (\underline{i} \blacktriangleleft \underline{e}_\perp).$$

Such undoing performs the *exclude* operation for the boundary case of a unit-element exclu*dee* already excluded from a generic exclu*dor*. That encourages us to think that it might work in general, like so:

$$\underline{dor} \text{ exclude } \underline{dee} \ = \ (\underline{dor} \blacktriangleleft {}^{!}\underline{dee}) \bullet {}^{\curvearrowleft !}\underline{dee}$$

... where the numeric dimension of the excludor is at least as big as that of its excludee. This does indeed work, and the reason is clear:

The excludor's extension by a unit-excludee reduces the separation of its higher-dimensioned result by the *sine* of the angle between them. Then, retraction re-creates the excludor at that reduced separation perpendicular to its excludee.

Again, such reducing re-creation should become clear by carefully working thru the details using the previous figures as an example:

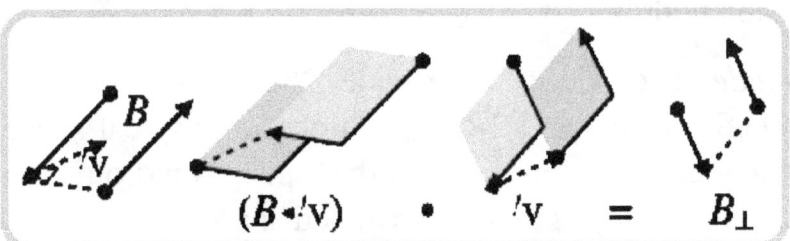

Exclude: a reducing extension, a re-creating retraction.

The excludor here, shown in perspective, is bivector B on the left; its excludee is vector v, normalized. Their preliminary extension in place gives their trivector result a volume separation reduced by the sine of the angle between them, as the slant between trivector ends makes clear.

This trivector is then squared-up against $'v$ in preparation for its geometric retraction by that unit vector. That re-creates the part of B perpendicular to v, as you see on the right. It is the same bivector exclusion shown previously.

Corroborating exercise: Deploy this version of *exclude* on the current number, $\underline{N} = i_1 \blacktriangleleft o + i_2 \blacktriangleleft o + i_1 \blacktriangleleft i_2 + i_2 \blacktriangleleft i_3$, to see if you get the same answer. You will need to calculate ...

$${}^2\underline{N} \; exclude \; {}^{[]}\underline{N} \;\; = \;\; ((i_1 \blacktriangleleft i_2 + i_2 \blacktriangleleft i_3) \; \blacktriangleleft \; {}'(i_1 + i_2)) \bullet {}^{\curvearrowleft}{}'(i_1 + i_2).$$

The success of this *exclude* is somewhat of a fluke—the ambient space here had enough elbow room to accommodate the initial extension. We will shortly see an example where that is not true, so the extension disappears prematurely via an attempt to ascend above the ceiling.

Include never has the problem of its initial *retraction* disappearing prematurely—the includee never has higher dimension than the includor; so the retraction never attempts to descend below the floor. Consequently, the *subtract-include* version of *exclude* is the only safe one. Nevertheless, the other one is lovely to contemplate as generalized extension undoing.

Include and *exclude* provide the tools we need to completely understand 2-compositions in physical space. Let us review and sharpen them on that space—and on a plane—in preparation for higher-dimensioned spaces.

Scenic route advisory: Readers traveling fast should be aware that we are now beginning compositional analysis that is primarily of concern beyond physical space, similar to the analysis Clifford indulged in when he introduced Clifford algebra. If that dismayed you (as it did me), your best strategy is to skip to the *Discovering Intersection* section, which doesn't require such mental rassling.

The following scenery will be of most value to readers crafting geometric-algebra programs and calculators. I have included it because I love such craftspeople.

Sharpening distillation tools

Since a bound vector *in any space* always initially coalesces to a sum of basis free vectors extended from the

209

origin, it always starts with the form v◄*o*. This is then translated by a parallel free bivector having the form v◄w, the free inclusion of v. This translator becomes algebraically removed from the free portion of the composition by coalescing with v◄*o* to form v◄(*o*+w). The free residue that remains *in any space* is the perpendicular exclusion of v.

The first space we shall analyze is a bound **3**-space, a plane; and the first sharpening we shall perform is to make the bound vector's normalized free part, $'v$, if present, the very first unit vector in our orthonormal basis, labeled plain i to distinguish it from the other basis vectors perpendicular to it.

In other words, we are starting with this special-purpose basis: *o* i i_2. The beauty of this analytic ploy is that an origin-bound vector becomes uniformly reduced to a mere scaled version of i◄*o*; and even more lovely, its parallel translator—the *inclusion*—becomes a scaled sum of free bivectors each having the form i◄i_j.

Hence, you can *read-off* the inclusion and exclusion just by looking at basis terms in the composition. This is a wonderfully clarifying simplification. See if you can find the inclusion for the two terms in this example: b(i◄*o*) and c(i◄i_2), the ceiling for this plane. ¿Is there an exclusion?

Answer: if the bound term b(i◄*o*) is not zero, then *the free term must be the translating inclusion* of it because that term has the form c(i◄i_2). The giveaway is the naked i in that extension. There is no exclusion here—no free term not containing i.

Entrenching the inclusion in the bound term generates i◄(*bo*+ci_2); a bound vector of dimension {**2**}. That leaves no residual free terms. If the bound term had been zero, the free term, $c(i \blacktriangleleft i_2)$ would have been left, a free bivector of dimension {2} (an effective exclusion because it is vacuously perpendicular to the absent bound term).

That exhausts the possibilities: A 2-composition in a plane can have only dimension **{2}** or dimension {2}, nothing else —a plane has no mixed 2-composition; no possibility of a free bivector perpendicular to a bound vector.

Altho the new algebra illuminating this is geometrically transparent here; its illumination is enlightening and will become crucial beyond physical space where transparency disappears.

Physical space is still transparent, but much richer. Under our new simplification it has primitive basis o i i_2 i_3, which generates the following generic bound-reduced 2-composition:

$$b(i \blacktriangleleft o) +$$
$$s_2(i \blacktriangleleft i_2) + s_3(i \blacktriangleleft i_3) +$$
$$r(i_2 \blacktriangleleft i_3) .$$

Notice that if this composition contains a bound vector (if b is not zero), then *two-thirds of the free composition is absorbed by that vector*, namely the parallel-on-i terms in the second row. They constitute the translating inclusion that gets entrenched within that vector. This is true even if that vector is left untranslated—these terms cannot be used for anything except bound-vector translation, if that vector is present.

(This fact may lose its surprise when you recall that the **entire** *free 1-composition* is effectively entrenched with the origin whenever it is not zero—**all** *of those terms* constitute the origin's translator, if it is present.)

Of course you might like to use the free 2-terms to express an arbitrary free bivector. In such a case you can't include a bound vector in its composition. In other words, if you want to express an arbitrary free bivector, you can't add a bound vector to it—you must stash them in distinct variables.

(Like the way you must stash a free vector and a point in distinct variables.)

Knowing all of that we can now completely characterize a 2-composition in physical space: If the bound vector is not zero, but its perpendicular residue $r(i_2 \triangleleft i_3)$ is, then the composition has dimension $\{2\}$. If that exclusion is not zero either, the dimension is $\{2, 2\}$. Finally, if the bound vector is zero, but not all of the free terms are, the dimension is $\{2\}$.

That latter singular dimension owes to vector overlap on any pair of basis bivectors, which immediately coalesces that pair to one free bivector. That bivector thence has a free-vector factor that is a scaled sum of the factors in the remaining term, so coalesces with it too. You saw this ploy at the beginning of this chapter, in a point-basis context.

2-compositions in higher spaces

Having sharpened our tools, let us move up one dimension to bound **5**-space. This space is no longer geometrically transparent, but it can still be made algebraically illuminating by using primitives o i i_2 i_3 i_4. They induce this *bound-reduced* generic 2-composition:

$$b(i \triangleleft o) +$$
$$s_2(i \triangleleft i_2) + s_3(i \triangleleft i_3) + s_4(i \triangleleft i_4) +$$
$$r_1(i_2 \triangleleft i_3) + r_2(i_2 \triangleleft i_4) + r_3(i_3 \triangleleft i_4) \, .$$

212

The main novelty now is that, if the bound vector $b(i \triangleleft o)$ is zero, then the purely free 2-terms that remain need not coalesce to a single bivector. For example, the sum of the first and last of them, $s_2(i \triangleleft i_2) + r_3(i_3 \triangleleft i_4)$, has no common extension factor by which they could coalesce; so this sum is irreducible—it contains two adamantly distinct free bivectors of mixed dimension $\{2, 2\}$

If the bound vector had not been zero, then half (only) of the free composition would become entrenched within it as its translator, namely the terms in the second row having i as a factor. That leaves the three free terms in the last row as perpendicular residue.

Those three terms each overlap on one primitive, so they always coalesce to one free bivector, as just explained. Hence, such a sum has dimension $\{2, 2\}$; unless there were no perpendicular terms, in which case it has dimension $\{2\}$

Knowing *all* of this, we can finally completely characterize 2-compositions in bound **n**-space for *any* **n**: If the composition does not contain a bound vector, that effectively removes the origin from the primitive basis, leaving $f = n-1$ free vectors. If f is even, these vectors can be extended pairwise without overlap to generate adamantly distinct free bivectors whose sum has dimension $\{2 \dots 2\}$ for $f/2$ distinct terms.

If f had been odd, the same pairwise non-overlapping extension could have been performed; but it would have left one primitive unpaired. That primitive, to generate a bivector, must be extended with a free vector in one of the previous pairs, so its extension will coalesce with that pair—it will not generate another adamantly distinct free bivector. Consequently, this sum will have dimension $\{2, \dots 2\}$ for $(f-$

1)/2 distinct terms.

These are the *maximum* dimensional distinctions possible for a purely free 2-composition in bound **n**-space; but of course most such will induce fewer distinctions. The individual terms within them will not be unique because they will depend on the coalescing tactic used. However, if these terms were coalesced minimally, then the mixed *dimension* generated will be unique.

For a 2-composition containing a bound vector, $b(i \triangleleft o)$, the maximum dimensional distinctions possible have a different form. Its translator has the form $s_2(i \triangleleft i_2) + s_3(i \triangleleft i_3) + \ldots$ which coalesces to $i \triangleleft v$, for $v = s_2 i_2 + s_3 i_3 + \ldots$ Hence, an existent bound vector in any 2-composition has the entrenched form $i \triangleleft (bo+v)$, a *single* bound vector, no matter how high **n** is. So, if there are no residual perpendicular free terms, such a composition has singular dimension {**2**} *in any space*.

If there are residual perpendicular free bivectors, they have no access to the origin *o*, of course; and they also have no access to the first basis vector, i, (under our illuminating analytic ploy) because it had been absorbed in the bound vector.

That leaves *n*-2 free primitives for the perpendicular part. They can be paired in a non-overlapping way, as previously done. The consequence is that this composition has slightly more maximum dimensional distinctions, but in a different form, like so:

A free 2-composition in a bound **n**-space has maximum dimension {*2, 2 ... 2*} for (*n-1*)/2 terms, where the <u>underscore</u> here is the *floor* operator, which effectively returns only the integer part of the division. Conversely, a bound 2-

composition in that space has maximum dimension $\{2, 2 \ldots 2\}$ for _(n-2)/2 free_ terms.

The preceding analysis is of primary significance for physical space, which can only support a non-composite twist. Beyond physical space the analysis merely shows that lines always sum to a single line, having possibly composite twist perpendicular to it. Such composition indicates that 2-dimensional perpendicularity in high dimensions requires distinct elements, but doesn't constrain how the basis primitives are distributed amongst them. Only in physical space is that unique.

Illuminating exercise: Show that the availability of the origin for a bound 2-composition sometimes allows it to have more elements than a corresponding free 2-composition. As an example, show that in **7**-space, a free composition has maximum distinction $\{2, 2, 2\}$; and a bound composition has maximum distinction $\{2, 2, 2\}$. However, in **8**-space, the max distinction for the free composition remains the same; but changes to $\{2, 2, 2, 2\}$ for the bound composition.

Distilling 3-compositions

In physical space, distilling 3-compositions is as trivial as distilling 2-compositions had been on a plane; and just as geometrically transparent:

In that space, a generic 3-composition has the possibility of a bivector bound from the origin plus a scaled version of the free-trivector ceiling. If the bound bivector exists, then the ceiling is certainly parallel to it; so it translates the bound bivector away from the origin via shape-shifting end-cancellation, leaving bound dimension $\{3\}$. If the bound bivector does not exist, that leaves only the trivector, free dimension {3}.

That exhausts the possibilities—physical space has no mixed 3-composition; no possibility of a free trivector perpendicular to a bound bivector.

Obvious tho this is geometrically, it is unfortunately extremely obscure algebraically under a fixed orthonormal basis. For such a basis, a bivector bound from the origin has this form: $b_1(i_1 \triangleleft i_2 \triangleleft o) + b_2(i_2 \triangleleft i_3 \triangleleft o) + b_3(i_3 \triangleleft i_1 \triangleleft o)$. The first step in coalescing this is to extract the origin: $(b_1(i_1 \triangleleft i_2) + b_2(i_2 \triangleleft i_3) + b_3(i_3 \triangleleft i_1)) \triangleleft o$

The next step is to coalesce the three free bivectors to one. You already know how to do this, but it is tedious and unenlightening; so I am going to do it for you so you can just gaze aghast: $(b_1 i_1 - b_2 i_3) \triangleleft (i_2 - b_3/b_2 i_1) \triangleleft o$.

(There is no possibility of primitive plurality here, at least syntactically, because these sums contain *primitive **elements***, already as free as possible.)

The final step is to translate this bound bivector by the scaled version of the ceiling $c(i_1 \triangleleft i_2 \triangleleft i_3)$. To do so you will need to factor the bivector's intricate free part into that trivector. Have fun with that—it is the algebraic way of shape-shifted end-cancellation, nearly always opaque.

Happily, for analytic purposes there is a way to sidestep all of this unenlightening tedium: reduce the bound terms in your 3-composition to *just one term*, as had been done for 2-compositions. The way to do that here is to make their free part a scaled extension of the first *two* vectors in your primitive orthonormal basis.

To distinguish these two bound-absorbed basis vectors from their unit-vector cohort, let us give them alphabetic subscripts: i_a i_b. Whence the bound bivector now has the singular form $b(i_a \blacktriangleleft i_b \blacktriangleleft o)$, and the free ceiling has the form $c(i_a \blacktriangleleft i_b \blacktriangleleft i_3)$

Clearly now, if the bound bivector exists (if b is not zero), then the free trivector is its translating inclusion, owing to their common bivector factor. So the bound bivector absorbs that trivector like so: $i_a \blacktriangleleft i_b \blacktriangleleft (bo + ci_3)$. That leaves no free terms, so this composition has bound dimension $\{3\}$. If the bound vector does not exist, that leaves only the trivector, free dimension $\{3\}$, as already discovered geometrically.

About this 3-poverty in physical space, the new agile bound-reduced algebra is *much* more transparent than the fixed-basis algebra had been. So we will deploy it immediately in the next-higher space, bound **5**-space—we shall give that space this primitive basis: o i_a i_a i_3 i_4. Its generic 3-composition then has this form:

$$b(i_a \blacktriangleleft i_b \blacktriangleleft o) +$$
$$s_3(i_a \blacktriangleleft i_b \blacktriangleleft i_3) + s_4(i_a \blacktriangleleft i_b \blacktriangleleft i_4) +$$
$$r_1(i_a \blacktriangleleft i_3 \blacktriangleleft i_4) + r_2(i_b \blacktriangleleft i_3 \blacktriangleleft i_4) .$$

Let us label this composition <u>N</u>, recycled. It is naive about the bound term, but we will soon fix that. Beginning naive is good because it avoids premature complexity.

The second row contains the translating inclusion containing the factor $i_a \blacktriangleleft i_b$. Those two terms simplify to $i_a \blacktriangleleft i_b \blacktriangleleft v$ where $v = s_3 i_3 + s_4 i_4$. This becomes entrenched within the bound part as $i_a \blacktriangleleft i_b \blacktriangleleft (bo + v)$, which is the translated bound bivector.

That leaves the third row as perpendicular free residue. It simplifies to $w ◂ i_3 ◂ i_4$, where $w = r_1 i_a + r_2 i_b$. Consequently, $\underline{N}$ has coalesced to $i_a ◂ i_b ◂ (bo+v) + w ◂ i_3 ◂ i_4$, dimension $\{3, 3\}$. If the perpendicular residue $w ◂ i_3 ◂ i_4$ had been absent, this would have been dimension $\{3\}$.

If the bound term had been zero, it could not have absorbed $i_a ◂ i_b ◂ v$, so $\underline{N}$ would have become $i_a ◂ i_b ◂ v + w ◂ i_3 ◂ i_4$. Analysis of that is surprisingly intricate (a foretaste of higher-dimensioned intricacy), as follows:

Since w is a scaled sum of i_a and i_b, it is a factor of $i_a ◂ i_b$; so the first term becomes $(w ◂ i_b ◂ v)/r_1$, as you may check. Similarly, the second term becomes $(w ◂ i_3 ◂ v)/s_4$. So, naively, their sum coalesces to $w ◂ (i_b/r_1 + i_3/s_4) ◂ v$, dimension $\{3\}$.

This is indeed true if *all* four free trivectors in the original basis exist. However if one term is absent, then the resulting sum of three trivectors is usually intrinsically composite, having dimension $\{3, 3\}$. Yes, that's right—four terms always coalesce to one here, but three terms usually coalesce to two. Intrepid readers might have fun rassling with this.

(Hint: an absent term means a zero basis coefficient. One consequence is ambiguous infinity if it is a divisor, as r_1 and s_4 are (under this particular coalescence). If it is not a divisor, as r_2 and s_3 are not, its vacuity induces a restrictive conflation of scalars. This took me some time to rassle down. Fortunately, none of this combinatorial intricacy is of much importance for the geometry, as shall become apparent.)

Now let us see why the bound term had been naive. Appreciating its naivety requires carefully verifying the preceding exclusion and inclusion, so here is a …

Corroborating exercise: Perform the inclusion and exclusion on the bound form of $\underline{N}$. Specifically, verify that $^3\underline{N}$ *include* $^{[I]}\underline{N}$ equals $i_a \triangleleft i_b \triangleleft v$ and $^3\underline{N}$ *exclude* $^{[I]}\underline{N}$ equals $w \triangleleft i_3 \triangleleft i_4$. Try both versions of *exclude*.

That exercise exposes a serious problem with the generalized-undoing kind of *exclude*. Here, that operation attempts to initially extend a free trivector by a free bivector in a free *4*-space,

... which exceeds the ceiling, so it vanishes; even tho the trivector automatically meets the criteria as an existent exclusion—it is already perpendicular to its smaller bivector excludee. That perpendicularity was manifested in its initial retraction during the *include* operation, which vanished. So the *subtract-include* kind of *exclude* is the only safe one.

That knowledge shall be important for exploring the naivety of $\underline{N}$'s bound term. It owes to the free *4*-expanse elbow-room in bound **5**-space, which allows the bound part of $\underline{N}$ to be a *mixed sum* $b_1(i_a \triangleleft i_b \triangleleft o) + b_2(i_3 \triangleleft i_4 \triangleleft o)$, dimension {**3**, **3**}; rather than a single element, $b(i_a \triangleleft i_b \triangleleft o)$, dimension {**3**} as had been naively presumed.

Such an irreducible bound composition has a peculiar delight: the inclusion for $i_a \triangleleft i_b$ is the exclusion for $i_3 \triangleleft i_4$, and vice-versa. You can just *read it off* of $\underline{N}$'s coalesced free terms: $i_a \triangleleft i_b \triangleleft v + w \triangleleft i_3 \triangleleft i_4$ (without even knowing what v and w are).

These contrary results might seem to make the inclusion of $i_a \triangleleft i_b + i_3 \triangleleft i_4$ contradictory; but in fact they make it *comprehensive*—the inclusion's translation of $\underline{N}$'s bound terms causes them to absorb all of the free terms. Let us explore this meticulously by explicitly calculating the

219

inclusion (even tho it can easily be read off):

Its inner operation is this retraction: $(i_a \blacktriangleleft i_b \blacktriangleleft v + w \blacktriangleleft i_3 \blacktriangleleft i_4) \bullet$ $(i_a \blacktriangleleft i_b + i_3 \blacktriangleleft i_4)$, which generates $-v - w$. That is then extended with the reverse of the includee: $(-v - w) \blacktriangleleft {}^{\hookleftarrow} (i_a \blacktriangleleft i_b + i_3 \blacktriangleleft i_4)$. This regenerates, (knowing that $v = s_3 i_3 + s_4 i_4$ and $w = r_1 i_a + r_2 i_b$), all of $\underline{N}$'s free terms, namely $v \blacktriangleleft i_a \blacktriangleleft i_b + w \blacktriangleleft i_3 \blacktriangleleft i_4$

These terms constitute the translator of the bound terms, $b_1(i_a \blacktriangleleft i_b \blacktriangleleft o) + b_2(i_3 \blacktriangleleft i_4 \blacktriangleleft o)$. On first thought this may seem strange because each term in the translatee is parallel to *only one term* in the translator; and perpendicular to the others. On second thought, that is precisely what is needed to translate a composite irreducible bound sum, like so:

The term that is parallel is absorbed in *just that one bound term*, and rejected by all the others because it is perpendicular to them. Consequently, translatee $b_1(i_a \blacktriangleleft i_b \blacktriangleleft o) + b_2(i_3 \blacktriangleleft i_4 \blacktriangleleft o)$ plus translator $v \blacktriangleleft i_a \blacktriangleleft i_b + w \blacktriangleleft i_3 \blacktriangleleft i_4$ produces …

$$i_a \blacktriangleleft i_b \blacktriangleleft (b_1 o + v) \quad + \quad i_3 \blacktriangleleft i_4 \blacktriangleleft (b_2 o + w)$$

You see that every free term in $\underline{N}$ has been absorbed by these intrinsically composite bound terms, leaving no residual perpendicular free terms. So this particular 3-composition has dimension $\{3, 3\}$.

In higher spaces, there might be residual free terms after the bound terms have absorbed their inclusion: $\{3 \dots \}$. Such terms constitute the perpendicular *exclusion* under the safe *subtract-include* version of the *exclude* operation. That, finally, exposes every possible scenario for …

Distilling a Mixed number M

Here are the preliminaries: The obvious first job is to partition $\underline{M}$'s basis composition into n-compositions of distinct n, since such compositions cannot coalesce with each other. Then trundle thru each one, coalescing it into distinct elements as much as possible, like so:

- If the composition contains bound basis terms (terms extended from o), then …

    1. Factor those terms into a sum of free $(n-1)$-vectors extended from o.

    2. Coalesce that free sum as much as possible. This will be the free *includee*, *dee*, needed for the *include* operation.

    3. Re-extend each *coalesced* term with the origin in preparation for translation to its confining space. This establishes the coalesced bound portion of $^n\underline{M}$ (coalesced $^n\underline{M}$), pre-translation, with dimension $\{\mathbf{n}...\}$

    4. Extract the free n-vectors in $^n\underline{M}$ using $^n\underline{M}$. This will be the *includor*, *dor*.

    5. Perform *dor* include *dee*. This will generate the *inclusion*, parallel to the coalesced bound terms.

    6. Coalesce that inclusion as much as possible; then use it to translate each coalesced bound term to its confining space, term-by-term, producing $^{Bn}\underline{M}$. This will remove the inclusion from the free terms and entrench it in the bound terms.

    7. The residual terms will be the free perpendicular *exclusion*, $^{Fn}\underline{M}$, under the safe *subtract-include* kind of *exclude*. They will have dimension $\{n...\}$ after they are coalesced.

8. That finally gives this n-composition dimension $\{\mathbf{n}.., n... \}$. That's it.

- If the composition does not contain bound basis terms, then ...
    1. Extract the free n-vectors using $^{n}\underline{M}$.
    2. Coalesce these n-vectors as much as possible, giving $^{Fn}\underline{M}$. This might be considered a kind of vacuous exclusion from a non-existent $^{Bn}\underline{M}$.
    3. That gives this n-composition dimension $\{n... \}$. That's it.

After you are finished distilling all of the n-compositions in $\underline{M}$, you will be left with with something whose minimal dimension will be $\{0, \mathbf{1}, 1, \mathbf{2}.., 2.., \mathbf{3}.., ... \}$ with omissions typically.

Such precise dimension constitutes a minimal form for $\underline{M}$. It is enlightening to us humans; especially for physical space where it makes crucial free–bound distinctions, and where it has coalesced to unique elements (unlike in higher spaces).

However, it will seldom be crucial for the algebra, which typically operates on the raw basis elements in a massively parallel, dimension-encompassing, bound-and-free way; without needing to coalesce their n-compositions at all. There is one important operation, however, for which coalescence *is* crucial, namely ...

Discovering intersection

Discovering intersection is an intrinsically **bound** operation that can be inherited by *free* elements via their *binding* extension from a point. (Conversely, discovering perpendicularity and parallelity are intrinsically *free*

222

operations that can be inherited by **bound** elements via *unbinding* them—extracting their free parts.)

The conventional free geometric algebra has no way within its algebra of articulating fixed intersection, so it typically resorts to an interpretation that tacitly anchors free vectors to an origin.

Under that interpretation, the infinite planes thru perpendicular free bivectors $i_1 \triangleleft i_2$ and $i_3 \triangleleft i_4$ are said to intersect *"in only one point"*; quite a surprise to young students. This is a true statement in the next space beyond physical space, but it is a curious one because the conventional algebra has no points available for intersection.

What is meant is that, since these bivectors have no common vector factor, the only thing their planes have in common is the tacit anchor point.

The full algebra does not have a tacit anchor point; it has an *explicit* one, and the intersection of these bivectors on it requires their extension from it, like so: $i \triangleleft i_2 \triangleleft o$ and $i_3 \triangleleft i_4 \triangleleft o$. Now they obviously do intersect on a **fixed point** because they have one as a common factor. But they don't intersect on a line because they don't have a **bound vector** as a common factor.

Such **bound** formalities are crucial for intersection—they precisely articulate ideas that are either absent in the conventional free algebra, or else very hazy owing to interpretations that are conceived differently by different people.

For example, ¿Have you ever read that perpendicular planes can sometimes intersect *nowhere at all*? (Like perpendicular lines can.) I haven't. But in fact the previous free bivectors can be bound in such a way that their infinite confining planes do not intersect *anywhere*: not on a point, not in a line, not in a plane—not anywhere.

¿In what space can that happen? Is it the same space in which they can intersect only on one point? Or does it require more elbow-room? Clear answers are coming up—*obvious* answers.

To arrive at them, let us begin with a simple example of planes that *do* intersect, and in more than just one point, namely these two bound bivectors:

$$B = i_1 \cdot i_2 \cdot o + 2\, i_2 \cdot i_3 \cdot o$$
$$C = 3\, i_2 \cdot i_3 \cdot o + i_3 \cdot i_1 \cdot o$$

Geometric intersection manifests itself algebraically as a common extension factor; so clearly these bivectors intersect on the origin, at least.

¿But do they intersect anywhere else? The only way to find out is to coalesce their basis compositions to single elements, whose factors can be checked for commonality. That is why basis distillation is crucial for intersection. Doing it generates …

$$B = i_2 \cdot (2i_3 - i_1) \cdot o$$
$$C = i_3 \cdot (i_1 - 3i_2) \cdot o$$

Inspection reveals no common factor other than the origin. However, we know that these bivectors can be shape-shifted *geometrically* into other factors. Indeed, if you could visualize them, you would have no trouble conjoining their shapes on their line of intersection. ¿Can that be done blindly —*algebraically*? Yes, by using the …

Fundamental factor-convergence tactic

The goal is to converge together an extension factor in one element to an extension factor in the other, hoping they become identical. The tactic is extremely simple, and obvious in retrospect. Simply augment one factor in an element by a scaled version of another factor in it:

$$u \blacktriangleleft v \;=\; (u + vv) \blacktriangleleft v \;=\; u \blacktriangleleft v + 0$$

This is algebraically obvious, and also geometrically transparent: it amounts to skewing one end of a bivector along its confining line.

To deploy this tactic, the general strategy is to use it over and over to walk a factor in one element to a factor in the other, alternatively; much like the way Euclid's greatest-common-divisor algorithm walks a factor in one number to a factor in the other, alternatively.

The factor-convergence strategy is more intricate, and proceeds by selecting from each element the most-composite unmatched factor in it. Then it converges those two factors if it can. If so, it repeats the process on remaining factors.

Choosing the most-composite factor has two benefits: first, it allows the less-composite factors to augment this one in small carefully-directed steps; and second, it precludes the annihilation that might ensue from augmenting a smaller factor by a larger one. Annihilation would happen if the summary terms in the larger factor contain every primitive in the smaller one.

Let us deploy the strategy on our example, B and C. Of the unmatched factors, the middle ones are most composite. For tidiness let us keep the primitive terms in those factors well ordered; so we begin with C whose second factor is already ordered. Then we walk B and C together like so:

$$C \;=\; i_3 \blacktriangleleft \; (i_1{-}3i_2) \;\blacktriangleleft o$$
$$B \;=\; -i_2 \blacktriangleleft \; (\overline{i}_1{-}\underline{2}i_3) \;\blacktriangleleft o$$
$$C \;=\; i_3 \blacktriangleleft \; (i_1{-}\underline{3}i_2{-}\overline{2i_3}) \;\blacktriangleleft o$$
$$B \;=\; -i_2 \blacktriangleleft \; (i_1{-}\overline{3i_2}{-}2i_3) \;\blacktriangleleft o$$

We have a match in just three simple steps. The goal for the first step was to match underlined $\underline{i}_1$ in C. That match, overlined, was done by negating B's first two factors. The goal for the second step was to match underlined $-\underline{2}i_3$. That match, overlined, was done by augmented the middle term of C by minus 2 times its first factor. The goal for the third step was to match underlined $-\underline{3}i_2$. That match, overlined, was done by augmenting the middle term of B by 3 times *its* first factor.

The algorithm has just discovered that B and C intersect on the confining line thru bound vector $(i_1{-}3i_2{-}2i_3)\blacktriangleleft o$. (This vector would typically be normalized to unit length to express directed locality, whose ±direction (orientation) would usually be ignored.)

With that success, the algorithm would now dutifully attempt to converge B and C on their first factor too. It would immediately notice that it could not do so because augmenting that factor in one element fails to converge it toward the same factor in the other.

Indeed, augmentation by the middle factor in either element annihilates that bivector entirely; augmentation by the third leaves its factors utterly unchanged. A well-crafted algorithm would notice this far better than most people. (Crafting such an algorithm might be an excellent programming exercise for

226

a college freshman, or a smart high-school student.) Here is what the three steps just done *look like*:

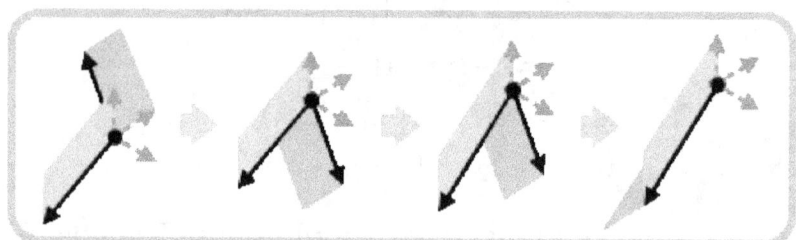

Shape-conjoining induced by factor-convergence

B is the darker bound bivector, **C** is the lighter one. Their direction (orientation) is not displayed here because it is not relevant to intersection. You see here that factor-convergence induces shape-conjoining. This intersection can also be computed by an apparently ...

More-elegant algebraic tactic

A person eager to use the elegant algebra of extension and retraction, rather than a successive approximation algorithm, might reason this way:

The orthogonal complements of these planes are vectors perpendicular to them that themselves can induce a plane by their extension. The orthogonal complement of *that* plane would be the intersection of the original planes. This idea is easy to visualize with some sketches.

Computing an orthogonal complement is done by retraction with the free ceiling, so this ploy is a purely free one. That means we are now ignoring the explicit origin; so we are back in the conventional algebra anchoring everything to a tacit origin. Here is how the algebra works for free B and C, having free ceiling $I = i_1 \triangleleft i_2 \triangleleft i_3$:

$$((B \bullet \mathrm{I}) \blacktriangleleft (C \bullet \mathrm{I})) \bullet \mathrm{I}$$

It almost looks like a music composition, and it proceeds like so: the two inner retractions generate $(-2i_1 - i_3) \blacktriangleleft (-3i_1 - i_2)$, as you may check. Their extension generates $2i_1 \blacktriangleleft i_2 - i_2 \blacktriangleleft i_3 + 3i_3 \blacktriangleleft i_1$. Its final retraction with the ceiling produces $-i_1 + 3i_2 + 2i_3$.

This is indeed the same intersection line that factor-convergence had produced. The *algebra* that arrived at it might truly be considered more elegant than a mere *algorithm*. But this algebra has serious problems.

First, as you may have noticed, it requires more computation: three retractive engagements with the metric and one extension. Factor-convergence, by contrast, makes no appeal at all to the metric; and it may well be finished with its work in the time it takes the first retraction to complete. Other problems are even more serious.

Most serious is that this algebra appeals obliquely to something that has no intrinsic relevance to the intersection between B and C, namely the free ceiling. ¿So how does that work when that ceiling changes to, say, the one in free 4-space?—orthogonal complements of bivectors there are not vectors.

The next problem this free algebra has is that it can't find fixed intersections of arbitrarily located elements. It either must presume that everything is anchored to the origin, so everything always intersects there; or else it must presume that intersections rove around. Different people embrace different presumptions.

The first presumption would imply that lines always intersect; the second would imply that they seldom do, at least in free *3*-space. The same applies for infinite planes in higher spaces; and even infinite volumes in still higher spaces; and so on up.

A final problem is that this algebra cannot compute intersection for elements of arbitrary dimension. It especially cannot compute intersection for elements of *different* dimension.

The factor-convergence algorithm has none of those problems: It is very efficient; it appeals only to the relation between the elements; it works in any space; it finds fixed intersections at arbitrary locations, of arbitrary dimension; and even different dimension. Let us peer at how it works for an …

Arbitrary-location example

We shall revisit the previous example; but with B and C translated to different confining planes:

$$B = i_2 \triangleleft (2i_3 - i_1) \triangleleft (o + 3i_1 + i_3)$$
$$C = i_3 \triangleleft (i_1 - 3i_2) \triangleleft (o + i_2 + 3i_3)$$

The translated origins here would have been generated by free trivectors in each basis composition parallel to B and C. We have already seen how that is done, so these bivectors are shown pre-translated.

You would naturally expect the algorithm to begin on the factors containing the origin because they are most composite. However, even if those factors had not been most composite, the algorithm would still begin with them because it must first establish a common anchor **point-factor** for the intersection. After finding that, it could start with the most-composite *vector-factor*, and go from there.

Degree of composition is a primary concern of the algorithm. It especially notices the completely non-composite i_2 factor in **B**, which can be used to match *any multiple* of that term in **C**'s point-factor; and it similarly notices the naked i_3 factor in **C** that can do the same for **B**'s point-factor. So it stashes those thoughts away until it can learn what multiples need to be matched.

It doesn't know at the outset because it realizes that **B**'s composite term $(2i_3-i_1)$ will need to augment its point-factor to match the $4i_3$ term in **C**'s point-factor; and **C**'s composite term (i_1-3i_2) will similarly be needed to match the $3i_1$ term in **B**'s point factor.

So it starts by using the composite factors as first-augmentors because in performing their job, they will induce scaling side effects in non-targeted terms that will need to be dealt with later.

¿How does the algorithm know all of this? It is because the clever programmer who crafted it made it scan all the factors in each element first and stash the coefficients of their terms in a table, ordered by degree of composition. Then the algorithm performs some intricate inter-factor analysis (that I shall side-step). Having done that, it can chug thru the intersection calculation mindlessly, but far better than you or I could mindfully.

Let's you and I walk thru what it would do, displayed here as a flow of convergence:

$$C = i_3 \blacktriangleleft (i_1-3i_2) \blacktriangleleft (o+i_2+\underline{3}i_3)$$
$$B = i_2 \blacktriangleleft (2i_3-i_1) \blacktriangleleft (o+3i_1+i_3 + (2i_3-i_1))$$

$$= \ldots \quad \blacktriangleleft \quad (o+\underline{2}i_1+\overline{3}i_3)$$
$$C \;=\; i_3 \blacktriangleleft (i_1-3i_2) \blacktriangleleft \;(o+i_2+3i_3 + 2(i_1-3i_2))$$
$$= \ldots \quad \blacktriangleleft \quad (o+\overline{2}i_1-\underline{5}i_2+3i_3)$$
$$B \;=\; i_2 \blacktriangleleft (2i_3-i_1) \blacktriangleleft \;(o+2i_1+3i_3 - 5(i_2))$$
$$= \ldots \quad \blacktriangleleft \quad (o+2i_1-\overline{5}i_2+3i_3)$$

We have a match as quickly as before, even tho this case was more intricate. That is the power of a well-crafted algorithm. Having scanned the factors and predetermined the order of augmentation, it focused on successively matching the <u>underlined</u> term in an above-bivector by properly augmenting the point-factor in a below-bivector, shown <u>overlined</u> as you see. It tells us that B and C intersect, at least, on the translated point $o+2i_1-5i_2+3i_3$.

With that success, the algorithm dutifully transfers attention to the unmatched most-composite vector factors, the middle ones. It has already seen that scenario before, and it knows that those factors converge to $(i_1-3i_2-2i_3)$.

So B and C intersect on the line thru the following bound vector: $(i_1-3i_2-2i_3) \blacktriangleleft (o+2i_1-5i_2+3i_3)$. This line is simply a translated version of the origin-bound line of intersection, as you might expect.

Different-dimension example

Now let us find the intersection of the previous bivector $B = i_2 \blacktriangleleft (2i_3-i_1) \blacktriangleleft (o+3i_1+i_3)$, dimension $\{3\}$; and a bound vector $v = (2i_1+2i_2) \blacktriangleleft (o+i_2+3i_3)$, dimension $\{2\}$. The algorithm begins, as emphasized, on their point-factors:

$$v \;=\; (2i_1+2i_2) \blacktriangleleft \;(o+i_2+\underline{3}i_3)$$
$$B \;=\; i_2 \blacktriangleleft (2i_3-i_1) \blacktriangleleft \;(o+3i_1+i_3+(2i_3-i_1))$$

$$= \ldots \blacktriangleleft (o+\underline{2}i_1+\overline{3}i_3)$$
$$\mathbf{v} = (2i_1+2i_2)\blacktriangleleft (o+i_2+3i_3+(2i_1+2i_2))$$
$$= \ldots \blacktriangleleft (o+\overline{2}i_1+\underline{3}i_2+3i_3)$$
$$\mathbf{B} = i_2\blacktriangleleft(2i_3-i_1)\blacktriangleleft (o+2i_1+3i_3+3(i_2))$$
$$= \ldots \blacktriangleleft (o+2i_1+\overline{3}i_2+3i_3)$$

The algorithm has just discovered that $\mathbf{v}$ and $\mathbf{B}$ intersect on point $(o+2i_1+3i_2+3i_3)$, so it turns its attention next to their most-composite vector-factors, namely $(2i_1+2i_2)$ and $(2i_3-i_1)$. Its inter-factor analysis has already informed it that the only augmentation available now is i_2 in $\mathbf{B}$'s factor; and that fails to provide a match on $\mathbf{v}$'s factor. So it concludes that these two elements intersect on only one point.

¿Did you notice that this different-dimension example proceeded in exactly the same way the same-dimension examples had? In fact, *the factor-convergence algorithm pays no **direct** attention to dimension at all*. It pays *indirect* attention simply by running out of factors to converge. So it can find the intersection of elements of arbitrary dimension.

Now we are finally ready to completely resolve the puzzle of ...

Non-parallel non-intersecting planes

Not only does factor-convergence pay no direct attention to dimension, it also pays no direct attention to ambient space either. This has interesting consequences for analysis of the intersection of previously mentioned free bivectors $i_1\blacktriangleleft i_2$ and $i_3\blacktriangleleft i_4$. For transparent exposition, let's label them $B_{1,2}$ and $B_{3,4}$.

They can't really intersect until they become bound;

meaning, among other things, that they acquire an extra extension factor. The algorithm certainly notices that, which causes it to indirectly notice ambient space, as follows.

The first thing it notices is that these two bivectors require free *4*-space, at least, because they require four different basis vectors. Their bondage therefor requires bound **5**-space, at least.

Speaking of bondage, the next thing the algorithm notices is that these bivectors can always be made to intersect on a point in that space—or *any* higher space—simply by binding them from the origin: $\boldsymbol{B}_{1,2} = i_1 \triangleleft i_2 \triangleleft \boldsymbol{o}$ and $\boldsymbol{B}_{3,4} = i_3 \triangleleft i_4 \triangleleft \boldsymbol{o}$.

(This is something the algorithm has known all along: *Any* two free elements whatsoever (and even scalars) can be made to intersect on a point, at least, by binding them from the origin.)

¿But can these planes be made to *not* intersect on a point by translating one of them away from the other? Suppose we try by translating $\boldsymbol{B}_{3,4}$ away from $\boldsymbol{B}_{1,2}$.

It is geometrically clear that we cannot hope to achieve non-intersection by using a vector parallel to $\boldsymbol{B}_{1,2}$ because that would just slide $\boldsymbol{B}_{3,4}$'s point of intersection to a different place on $\boldsymbol{B}_{1,2}$'s plane.

Let us corroborate that algebraically. Suppose $\boldsymbol{B}_{3,4}$ has been translated like so: $i_3 \triangleleft i_4 \triangleleft (\boldsymbol{o}+s_1 i_1+s_2 i_2)$. The factor-converging algorithm can match $\boldsymbol{B}_{1,2} = i_1 \triangleleft i_2 \triangleleft \boldsymbol{o}$ to that trivially, simply by augmenting its point-factor by s_1 times its first factor and s_2 times its second. Whence these two bivectors intersect on point $(\boldsymbol{o}+s_1 i_1+s_2 i_2)$ after the translation.

233

So clearly we must translate $B_{3,4}$ perpendicular to $B_{1,2}$ to have any hope of them not intersecting. That means that we must use this translated form of $B_{3,4}$: $i_3 \triangleleft i_4 \triangleleft (o + s_3 i_3 + s_4 i_4)$.

Now we have a reverse triviality: The factor-converging algorithm can match *that* to $B_{1,2} = i_1 \triangleleft i_2 \triangleleft o$ simply by "augmenting" its point-factor by $-s_3$ times its first factor and $-s_4$ times its second. Whence these two bivectors still intersect on the origin after the translation.

This is a novel kind of factor-convergence, namely *term-elimination* convergence. It was not considered in the previous examples; but if it had been, a previous anchor point could have been made less composite, a challenge for intrepid readers.

So these two bivectors always intersect on a point. You can make them intersect on *any* point you want, simply by translating their point-factors by the same free vector.

The irrepressible point-intersection of these two bivector is impossible to visualize, but there is a corresponding example that can be visualized: Two perpendicular planes in physical space have a line of intersection. If one plane is moved along another, they still intersect, but in different lines. If one plane is moved perpendicular to another, they still intersect in the same line. They have irrepressible line-intersection that can be moved anywhere you want.

From the analysis so far, perpendicular planes in higher spaces apparently always intersect in a point, at least. But wait, the reason the *perpendicular* translation failed was that it was actually *parallel* to itself. ¿What if we were able to translate $B_{3,4}$ perpendicular to the other bivector *and* itself?

That cannot be done in bound **5**-space because that space has no vectors perpendicular to both $B_{1,2}$ and $B_{3,4}$. But bound **6**-space does, namely i_5. So let us move up to there and try this translated form of $B_{3,4}$: $i_3 \blacktriangleleft i_4 \blacktriangleleft (o+i_5)$.

Now the factor-convergence algorithm is stymied—neither bivector has an i_5 factor that can converge on $(o+i_5)$. So we have finally generated two planes that do not intersect *anywhere*:   not on a point, not in a line, not in a plane—not anywhere.

This has become algebraically obvious even tho it is impossible to visualize geometrically. So planes in high spaces can indeed intersect on only one point, or even none at all, like lines in physical space.

Synthesis

This chapter attempts to simplify and unify the path taken so far, which culminates about six millennia of human numeric meandering. The journey began in prehistory with the rules of addition, which shall play a starring role as cohering heroes in this chapter:

$$\bar{a} + \bar{b} = \bar{b} + \bar{a}$$
$$(\bar{a} + \bar{b}) + \bar{c} = \bar{a} + (\bar{b} + \bar{c})$$
$$\bar{a} + 0 = \bar{a}$$
$$\bar{a} + (-\bar{a}) = 0$$

These rules were only codified like this a little over a century ago; but shells used as coins (inland of course, in fixed supply) indicate that they were understood, in *very* naive form, beginning around 4,000 BCE, perhaps earlier. The over-bars are intended to suggest how these numbers were almost certainly understood at that time: as primordial *adding numbers*, meaning integers. They naturally induced the inverse operation of subtraction, whose undoing consequence is formalized in the last rule.

Multiplying numbers likely arose after adding numbers began to be used extensively, but they took well over a thousand years to make an explicit appearance, at the onset of history (we humans are slow about math). They evidently started with multiple-addition of, say, 5 bunches of 7 bananas; and then progressed to how much land is in a garden of perhaps 3 by 9 standard knots of rope.

Their inverse operation, division, generated fractions like 1½ by 4½ knots of land, meaning 27/4 squares of it; so multiplying numbers are the rational ones. They obey these rules, which shall play a starring role as imitative bunglers:

$$a \times b = b \times a$$
$$(a \times b) \times c = a \times (b \times c)$$
$$a \times 1 = a$$
$$a \times (/a) = 1$$

The mid-bars are intended to suggest that these numbers articulate fractions; and again, their rules were only recently codified in this form. They encompass adding numbers, an expressive ancient unification. But then we humans did not advance beyond combined *adding-multiplying numbers* for another two thousand years or so (*very* slow we are about math).

Finally, around 500 BCE, the *scaling numbers*—the so-called *real* ones—arose with the startling discovery that fractions cannot represent all numbers, no matter how finely they divide. These new numbers encompass the old adding-multiplying ones, another ancient unification. They did not induce further rules; but their upset of naive conceptions was humankind's first major scientific shock.

That shock persisted for centuries until Euclid finally devoted about one third of his work grappling with it. The puzzlement lingered even into the Common Era, but eventually faded and became overshadowed by the subsequent *earth-is-not-the-center* shock, the *humans-and-crabs-have-the-same-grandparents* shock, the *time-depends-on-speed* shock, and our present *quantum* shock.

237

Notice the seductive symmetry between the rules for addition and multiplication. That symmetry—from a spatial-arithmetic perspective—is specious, a deception, a hindrance, a trap. It has kept us stuck with dimensionless *adding-multiplying-scaling numbers* for the past several thousand years.

For example, weighted points arose nearly two centuries ago, twice—independently; and it was realized that they unexpectedly obey the same rules of addition that numbers obey. However, they were reflexively rejected as numbers because their multiplication does not obey the commutative rule that numbers obey.

We will have finally begun to advance beyond antediluvian dimensionless numbers when middle-school students understand how to add dimensioned point numbers. We will have truly got underway when high-school students understand how to multiply them to higher dimensions.

The lovely symmetry between the previous rules is an expressive simplification when it is genuine; but an intellectual impediment when it is not. Unfortunately, it is not genuine even in scalar arithmetic, as we shall ponder shortly; and especially not in well-dimensioned spatial arithmetic. To prepare for that understanding, let us peek first at some …

Genuine symmetry

Look at the elegant rules for the algebra of sets, usually denoted with capital letters; with $\cup$ indicating union, $\cap$ indicating intersection:

$$A \cup B = B \cup A$$
$$(A \cup B) \cup C = A \cup (B \cup C)$$

$$A \cup \textbf{none} = A \quad .$$
$$A \cup \overline{A} = \textbf{all}$$

$$A \cap B = B \cap A$$
$$(A \cap B) \cap C = A \cap (B \cap C)$$
$$A \cap \textbf{all} = A \quad .$$
$$A \cap \overline{A} = \textbf{none}$$

Notice the complete symmetry here: simply interchange union $\cup$ and intersection $\cap$, **none** and **all**, in one set of rules and you get the other set. (An over-bar indicates *set complement*, meaning all elements outside that set.) Even the interactions between these two groups of rules has complete symmetry:

$$A \cup (B \cap C) = (A \cup B) \cap (A \cup C)$$
$$A \cap (B \cup C) = (A \cap B) \cup (A \cap C)$$

All of this symmetry allows the rules for intersection to be derived from the rules for union; so a person might suppose that the rules for multiplication could similarly be derived from the rules for addition.

Historically that has been partly true because, until recently, multiplication has always been considered *multiple-addition*. In consequence, it inherited addition's commutative and associative laws.

So it seemed only natural and proper that it should inherit addition's identity and inverse laws too, as it appears to have done. After these four rules became frozen-in, multiplication was hindered from advancing to dimensioned numbers.

The freezing-in of the rules of *addition*, by contrast, did not hinder *that* operation from advancing to dimensioned numbers, or indeed *any* kind of numbers at all. To say this differently, the freezing of the rules of addition represent a proper finalization: a conclusive encoding of *universal* meaning within syntax. It exemplifies the very best kind of ...

Facilitating symmetry

To demonstrate the universal nature of the rules of addition, let us rewrite them in terms of the Mixed Numbers this book has been grappling with:

$$\underline{M} + \underline{N} = \underline{N} + \underline{M}$$
$$(\underline{M} + \underline{N}) + \underline{P} = \underline{M} + (\underline{N} + \underline{P})$$
$$\underline{M} + 0 = \underline{M}$$
$$\underline{M} + (-\underline{M}) = 0$$

These rules encompass the adding-multiplying-scaling numbers, a unification that is only just now getting underway. Having put ourselves at the leading edge of glacial mathematical progress, let us now go back six thousand years to understand how *very* naively these rules were understood at first, in prehistoric terms.

The first rule says that whether you pay me three shells now, and then four later; or four now, and then three later, you will have paid me the same number of shells, namely seven.

The second rule says that whether you pay me three and four shells now, and then five later; or three now, and then four and five later, you will again have paid me the same number, namely one dozen.

The third rule, in prehistoric terms, says that if you pay me some shells, and then no more later, I am stuck with what you first paid me.

The final rule, also in very prehistoric terms, says that if you pay me some shells, and then take them all away later, you will have paid me nothing.

"*Well duh*", a prehistoric person might say, "*none of this requires any thought.*" True, and that is why the rules were only recently encoded—*it was finally realized that explicit rules could enable thoughtless thinking*; could make intricate calculations automatic.

It was also realized—after addition's rules were finally encoded—that even the most obvious of them might be violated by other operations; even multiplication, strange tho that seemed: order might matter, grouping might matter.

For rules to participate in thoughtless thinking, they need symbols they can manipulate; so humankind's first enabling task was to come up with number symbols. Various encodings arose at the onset of history, but the symbol for *Nothing* in the last two rules then took a surprisingly long time to appear— many thousands of years in fact.

It first appeared around 500 CE, and then took a few more centuries to become assimilated. At first it was not considered a number, but merely a handy symbol for *Nothing* in arithmetic. (It would have been just as handy in set algebra as **none**, but that idiom was more than a millennia in the future.)

Its handiness in arithmetic eventually caused it to be viewed as just another number—another *scaling* one, another *real* one, the only kind available in arithmetic at that time. (Or now.)

That works fine for the rules of addition, but causes problems for the rules of multiplication, as we shall see shortly. In preparation, let us look at how it works for addition.

0 was the second *do-nothing* symbol humankind ever came up with. The very first one, namely the various forms of 1, had already been doing something in the adding numbers. Only when it was promoted to the multiplying numbers did it begin to do nothing. Ha. In other words, 1 was a *fortuitous* do-nothing symbol.

0, by contrast, was an *intentional* do-nothing symbol, an empty box, specifically intended at first to do nothing in addition. The realization that its doing nothing would be useful was a watershed moment in mathematics — it finally enabled symmetry to be specified by rules.

Symmetry, in essence, is a change that causes no change — a change in something that induces no change in something else. The unchanged part is usually considered more important than the changed part (which is often not even apparent, as in bilateral symmetry). The symmetry between the set rules, for example, allows you to exchange union with intersection, **none** with **all**, without causing a change in the validity of the rules.

Each individual rule, in fact, is itself a symmetry: a change from its left side to its right side, or vice versa, causes no change in its equality.

The significance of this for a *do-nothing* symbol like 0 is that it encodes the most important part of a symmetry, the *no-change* part. In recognition of that significance, do-nothing symbols came to be called *identities* for their operations: 0 is the identity for addition; 1 is the identity for multiplication.

The rules for addition are essentially a bag of symmetries, a sophisticated idea that dawned on humankind only recently.

242

But these particular rules constitute a *very special* bag because the last half of them encodes the crucial *no-change* part of a symmetry: the identity rule, just described, and the *inverse* rule.

The *inverse* of an element is the unique element that, when combined with it, generates the identity. Said more transparently, an element combined with its inverse causes *no change*. For addition, the inverse of a number is its *negative*, encoded in the last rule.

These *no-change* ideas only became obvious after the rules for addition had finally been encoded. Gazing at them, a few clever mathematicians realized they can be simplified and abstracted from numbers to express the generic idea of symmetry.

Thus was born the idea of a formal *group*: an arbitrary operation on a set of elements that obeys the last three laws of addition. If the set obeys the first law too, it constitutes a *commutative group*. Groups express symmetry; scaling numbers express quantity.

Having arrived at that insight, mathematicians then turned the idea on its head—they began to declare that addition is merely a commutative group with identity 0, and negatives as inverses. This is truly an elegant advance, but it subsequently went too far.

Hampering symmetry

It would seem that multiplication is also a commutative group; but with identity 1, and reciprocals as inverses. So generic numbers would seem to be simply those things that obey the rules for both the addition group and the multiplication group. Right?

Right! according to conventional wisdom. Such generic elements constitute a *field*, widely venerated, widely deployed.

There is a flaw with that idea: not all numbers have reciprocals, so multiplication does not really constitute a group. Fortunately the flaw is *very* tiny for dimensionless scaling numbers, and arose from good intentions: the desire to promote to multiplication the do-nothing symbol 0 that had initially been intended for addition.

Understanding the consequences of its promotion requires first specifying multiplication's relation with addition:

$$a \times (b + c) \;=\; a \times b \;+\; a \times c$$

This is usually called multiplication's *distributive rule*, based on the way it grammatically distributes across addition. However, it would be more meaningfully called multiplication's *respect for summary* (I think I mentioned this before): the product of a sum is a sum of products. It was obvious from the dawn of history because multiple-addition automatically enforces it.

(This is especially obvious with integers—try a sketch. This property is now often obscurely termed *linearity*, a term that seems to universally mystify pedestrians and young children, like *entropy* does.)

Notice that this respect is a one-way street, unlike in the rules for sets: summary does not similarly respect multiplication like union does similarly respect intersection. This is our first hint that the conventional field-axioms' presumption of elegant symmetry between the rules of addition and multiplication might be mistaken.

Our second hint arises when we deploy $0 = b+(-b)$ on multiplication's respect for summary:

$$a \times 0 = a \times (b + -b) =$$
$$a \times b + -(a \times b) = 0$$

In other words 0 times anything is 0, exactly what was expected when *Nothing* was first promoted to multiplication. What was not expected at first was that this precludes 0 from having a reciprocal: there is no number that can possibly multiply 0 to the identity 1.

When that idea finally dawned on humankind a millennia ago or so it was expressed more mysteriously, more authoritatively, more patronizingly: *"You can't divide by zero."*

It is *still* expressed that way, a dismaying fact that indicates humankind has not yet assimilated the idea that division is merely multiplication by a reciprocal (in the same way that subtraction is merely addition of a negative). Look again at the final rules for multiplication and addition:

$$a \times (/a) = 1$$
$$a + (-a) = 0$$

Each of these rules is expressed in terms of an *inverse*, namely multiplication of *reciprocal* /a, and addition of *negative* -a. There is excellent reason for deploying inverse *elements* like this, rather than their inverse *operations*: multiplication and addition have simple elegant symmetries, previously displayed; but division and subtraction do not.

Mathematicians understand this in their bones, but the expressiveness of the idea has not yet seeped outside their domain. And even mathematicians, who have long used the *addition-of-a-negative* notation for subtraction, are only just now beginning to use the *multiplication-by-a-reciprocal* notation for division.

(By the way, I am not a mathematician, but I do like that tribe. I am a Groucho—I would never belong to any group that would have me as a member.)

That notation is crucial for advance to well-dimensioned numbers. It is key to the tiny crack in the field axioms for multiplication of dimensionless numbers: /0 does not exist—zero has no reciprocal. That crack shall widen into a chasm when humankind eventually advances to dimensioned numbers.

Since zero has no multiplicative inverse (¿does it have an additive one?), multiplication does not really constitute a group, like addition does, altho it *almost* does. There are two possible responses to this problem.

Here is the first: Since the problem seems so tiny, it can be patched over simply by asserting that multiplication really does constitute a commutative group, provided you meticulously exclude zero. This has been the historical response during the last century, manifested in hundreds of different textbooks.

The other possible response is that maybe this tiny crack is an indicator of a bigger problem: ¿Maybe zero, useful tho it undoubtedly is, is not really a number? Maybe there are numbers, *very expressive numbers*, that do not have reciprocals either? If so, maybe the field axioms for multiplication would better be replaced by more-expressive, but less-elegant rules? Rules that finally abandon their

246

historical faux symmetry with the rules of addition?

All of that is true; and the raw material for such expressive rules is …

The articulate kludge so far

Semantic axioms, full algebra

Scalars have dimension {0}
0 has dimension {}
Bound primitives are points, dimension {**1**}

Free primitives are free vectors, dimension {1}, meaning dimension {**1, 1**}-*without magnitude*.

An extension of n primitives, of which at least one is a point, is a bound n-element, dimension {**n**}. It has spatial expanse of n-1; so it is called a bound **n–1**-vector.

An extension of n free vectors is a free n-element, dimension {**n**}, meaning dimension {**n**, **n**}-*without magnitude*. It has spatial expanse of n, and is called a free n-vector.

The dimension of a mixed number $\underline{M}$ is specified by the collective dimensions of the elements in its sum, when distilled minimally.

The *magnitude* of an element is the separation of its free part. Since the free part of a free element vanishes, its magnitude also vanishes.

Syntactic axioms, full algebra

Addition

$$\underline{M} + \underline{N} = \underline{N} + \underline{M}$$
$$(\underline{M} + \underline{N}) + \underline{P} = \underline{M} + (\underline{N} + \underline{P})$$
$$\underline{M} + 0 = \underline{M}$$
$$\underline{M} + (-\underline{M}) = 0$$

Scalar multiplication

$$a\,\underline{M} = \underline{M}\,a$$
$$(a\,b)\,\underline{M} = a\,(b\,\underline{M})$$
$$\underline{M}\,1 = \underline{M}$$
$$a\,(/a) = 1$$
$$a(\underline{M} + \underline{N}) = a\,\underline{M} + a\,\underline{N}$$

Extension

$$\underline{e}_{odd} \blacktriangleleft p = -(p \blacktriangleleft \underline{e}_{odd})$$
$$\underline{e}_{even} \blacktriangleleft p = p \blacktriangleleft \underline{e}_{even}$$
$$(\underline{M} \blacktriangleleft \underline{N}) \blacktriangleleft \underline{P} = \underline{M} \blacktriangleleft (\underline{N} \blacktriangleleft \underline{P})$$
$$\underline{M} \blacktriangleleft 1 = \underline{M}$$
(So $\underline{M} \blacktriangleleft a = \underline{M}\,a$ via
scalar multiplication.)
$$(\underline{M} + \underline{N}) \blacktriangleleft \underline{P} = \underline{M} \blacktriangleleft \underline{P} + \underline{N} \blacktriangleleft \underline{P}$$
(So a free vector, as a sum of points,
commutes like points do.)

Free-part extraction

$$^{[]}(\underline{e} \blacktriangleleft p) = \underline{e}$$
$$^{[]}(\underline{M} + \underline{N}) = {}^{[]}\underline{M} + {}^{[]}\underline{N}$$
(So the free part of a free element
vanishes because that is the sum
of opposed bound elements.)

Semantic axioms, free sub-algebra

Retraction of a free element with a free vector produces a result with decremented dimension, if it does not vanish.

Retraction of two free elements is iterated retraction by the free-vector factors in the smaller-or-same-dimensioned element.

The *separation* of a free element $\underline{e}$ is $\sqrt{|\underline{e} \bullet \underline{e}|}$.

Syntactic axioms, free sub-algebra

Retraction

$$\underline{e}_{odd} \bullet v = v \bullet \underline{e}_{odd}$$
$$\underline{e}_{even} \bullet v = -(v \bullet \underline{e}_{even})$$
$$\underline{e} \bullet /\underline{e} = 1$$
$$\underline{M} \bullet a = 0$$
$$(\underline{M} + \underline{N}) \bullet \underline{P} = \underline{M} \bullet \underline{P} + \underline{N} \bullet \underline{P}$$

This is a rambling hodge-podge of rules; but a very expressive one, in the same way that a natural language is a very expressive hodge-podge. These rules contain three kinds of multiplication; which, amongst themselves, violate the field axioms in almost every way possible, like so:

Neither extension nor retraction always commute. Extension is associative but retraction is often not. Conversely, every free element $\underline{e}$ has a retraction inverse, $\underline{e}/\underline{e} \bullet \underline{e}$, but not an extension inverse. Extension has an identity but retraction does not. Nevertheless, there are two genuflections to the field axioms:

The first is that multiplication of pure scalars now truly is a commutative group because 0 is not a scalar. In fact it is not even a number, but rather *no thing* with *no properties* whatsoever.

(Ironically, that expressive distinction collaterally invalidates addition of *pure scalars* as a group because it removes the additive identity from the scalars, which also removes addition's inverse axiom—the field axioms require richer distinctions when numbers become well-dimensioned.)

The second genuflection to the field axioms is more of a deep bow to the one rule they get exactly right: All three of these products respect summary (so that constitutes *the* distinguishing property of a genuine product).

That fact, combined with the complementary commuting properties of extension and retraction, and the complementary lacunas in their axioms, gives hope that they might be unified somehow. Clifford did that, but only for the free sub-algebra (altho he was likely unaware of that restriction—I have been unable to find mention of it in his writings).

Unifying the free sub-algebra

Recall that Grassmann was motivated to define an inner product by completeness, like so: his *outer* product of an element with a vector pays attention only to the part of the vector perpendicular to the element, the part *completely outside* it; so he began to define an *inner* product to pay attention to the other part—the part parallel to the element, the part *completely inside* it.

After he succeeded, you might guess that he may have tried to synthesize these two products into a composite sum that pays attention to the entire vector, its rejection and projection both.

That idea may have crossed his fertile young mind; but if so it was promptly rejected because that mind had decided

250

right at the outset that *"Any two magnitudes of the same order, but only such can be added."*p290

Late in life he saw this apparently obvious apples-vs-oranges principle violated in an unexpectedly fruitful way—he happened to read an article extolling the virtues of Hamilton's quaternion product, a composite sum of his *Scalar* and *Vector*.

That article changed Grassmann's still-supple mind about what can be added together. We know it did because it provoked him to publish an aggrieved huff in 1877, the last year of his life, claiming a kind of *latent* priority for Hamilton's idea. Its title, *The Position of the Hamiltonian Quaternions in Extension Theory*, presages its somewhat meandering introduction, which is worth replicating in its entirety to demonstrate its vigorous indignation:

> Since extension theory makes only *one* arbitrary assumption, that is that there exist magnitudes that can be numerically derived from more than one unit, and proceeds from this in a completely objective way, all expressions that are numerically derivable from a number of independent units ["*displacements*" was what he had in mind here], and in particular the Hamiltonian quaternions, have their definition in extension theory and only find their scientific foundation in it. This was not previously recognized, and Goran Dillner, in his instructive article on quaternions does not mention extension theory even once, although he derives a whole series of theorems from the theory of quaternions which have already found their basis much more simply and emerge more naturally from the nature of the subject in my *Ausdehnungslehre* of 1844, and likewise in the later revision of 1862. Also it is reprehensible, and little serviceable to the theory of quaternions, that according to Hamilton's procedure one designates simple and long familiar concepts by new, often unsuitable terms, as "vector" for "displacement", "tensor" for "length" or "numerical value", etc.p525

For balance, it is only fair to glance at Hamilton's

complementary misgivings about Grassmann, sent to De Morgan in 1853 in a letter. Its jocular criticism is somewhat complimentary, at least tacitly:

> His *outer* products I think that I do understand, and that is saying something for a person who has not learned to smoke. And even his *inner* products, published subsequently to the *outer* ones (in 1847), I can swallow pretty well. In fact, the "inner products" of Grassmann have much analogy to my "*scalar parts*" of a quaternion, and his "outer products" to my "*vector parts*". If the notion of *combining* them had occurred to him, he *might* have been led to the quaternions; but those he seems to me to have altogether failed to perceive. [*A History of Vector Analysis*, Crowe, p86.]

The notion of *combining* them—according to Grassmann's introduction just replicated—was implicit in his mind *some thirty years retroactively*. To demonstrate that, he proceeded to display a "*product ab of two displacements* [whose essence in the notation of this book] *can be represented in the form*"

$$ab = a \bullet b + a \blacktriangleleft b$$

He called this the *central product* because "*it forms the central order between the two principal types of multiplication that I have called 'outer' and 'inner'*." Which is to say, Grassmann now conceived this new product to have been latent in his original products, ex postfacto.

But this product is not quite right for the quaternion product of a and b, which Hamilton would express as ab = S.ab + V.ab. Its *scalar part*, S.ab, has the opposite sign to the modern dot product (because both Gibbs and Heaviside independently reversed it for comprehensibility); and its *vector part*, V.ab, is not a bivector. Grassmann patched all of that by negating the scalar and taking the complement of the bivector, like so:

$$ab = -a \bullet b + |(a \blacktriangleleft b)$$

Whence *"follow all the rule of quaternions, and indeed most of them with the greatest of ease"* he claimed. This demonstrates that Grassmann, even in his old age, was able to achieve a better understanding of quaternions than their own creator had.

In particular, it shows that he recognized Hamilton's vectors as bivectors in disguise, a ploy that only works in physical space where vectors and bivectors can be orthogonal complements of each other. This idea does indeed *"emerge more naturally"* from Grassmann's algebra than from Hamilton's, where it doesn't emerge at all.

However, Grassmann neglected to mention that his latent heretofore-unseen central product abandons his long-standing denial of composite sums. He also neglected to mention that in making it associative, he was combining a non-associative product, a•b, with an associative one, a◄b. That very useful idea turns out to be unexpectedly valid, but it requires some scrutiny to understand why, as we shall see.

Grassmann's *geometric synthesis* was anticipated by a now widely celebrated *algebraic analysis* that William Clifford presented to the London Mathematical Society in 1876. It was published posthumously in Clifford's 1882 *Mathematical Papers* under the title *On the Classifications of Geometric Algebras*. The introduction to that book declared that *"Clifford was above all and before all a geometer."*

Indeed, he had devoured all the available polyglot literature on geometry including Mobius's *Der Barycentrische Calcul*, Argand's and Gauss's planar representations of complex numbers, most of Gauss's other work, Hamilton's *Quaternions*, Grassmann's two books on *Ausdehnungslehre*, Cauchy's *Clefs Algebiques*, most of Maxwell's geometric ideas, and all the other shorter and less luminous works he could get his eager young hands on.

He began to play with these various "*geometric algebras*" as he called them, and discovered an astonishingly simple and fertile way to unify them: he simply bolted Hamilton's scalar product of *vectors* onto Grassmann's outer product of *displacements*, like so:

> Physical considerations [extraction of length] however lead us to regard [Hamilton's] i^2 as a scalar even when i is regarded as a vector [rather than a quarter-turning operation]. For these purposes it does not matter whether it is put to -1 or $+1$. [It would matter for a quarter-turning operation, whose square is a half turn, namely -1.] I propose here to extend this assumption to the Grassmann representation in general: i.e., I take n units $i_1\ i_2\ \dots\ i_n$ such that $i_s^2 = +1$, and $i_r i_s = -i_s i_r$.

Since Clifford here decided to make all of Grassmann's i_s basis vectors square to $+1$, like Grassmann's inner product did, rather than -1 like Hamilton's scalar product did, he had really bolted Grassmann's *own* inner product onto his outer product.

However, he could not have left Hamilton out of the picture because it was Hamilton's juxtaposed *commuting–neg-commuting multiplication* that motivated his idea. Clifford was simplifying and enhancing Grassmann's bracketed *neg-commuting multiplication* with Hamilton's juxtaposed *commuting scalar-product* properties.

At first glance this ploy seems bizarre and contradictory. Clifford's new product asserts that $i_r i_s = -i_s i_r$, a rule that derives from Grassmann's extension: $[i_r i_s] = -[i_s i_r]$, stripped of brackets. But when extension is applied to $i_s^2 = i_s \blacktriangleleft i_s$, it becomes 0 since it equals its own negative. And yet Clifford is claiming it is 1. ¿Isn't that a contradiction?

It would be if $i_s \triangleleft i_s$ truly *becomes a number*. Fortunately, it doesn't—*it becomes nothing; it vanishes; it gets out of the way* so that i_s^2 can become $i_s \bullet i_s$, which *does* equal 1, at least in Grassmann's basis.

In other words, Clifford's new juxtaposed product $i_r i_s$ implicitly comprises the following *composite* product: $i_r \bullet i_s + i_r \triangleleft i_s$. When $_r$ equals $_s$, the outer product gets out of the way so the inner product can take over—that product truly was bolted on.

For this attachment to work when $_r$ *does not* equal $_s$, the inner product must conversely get out of the way so the outer product can take over. This can only happen if all the i_s basis vectors are perpendicular to each other.

In other words, Clifford presumed a basis having an orthonormal metric like most everyone used at that time, a fact he did not mention. In any case, his focus in this paper was on algebraic intricacies exclusively, rather than geometric meaning.

Focus on geometric meaning provokes this question: ¿Does Clifford's new product separate into Grassmann's latent central product ab = a•b + a ◂ b for *arbitrary* (non-basis) vectors a and b, as it seems to?

Yes, and it is trivial: when a and b are expressed as scaled sums of the i_s basis vectors, Clifford's product of them generates terms in which $_r$ does equal $_s$, and terms in which $_r$ does *not* equal $_s$. The former terms comprise the retraction a•b; the latter terms comprise the extension a ◂ b. Let's look at an example, reverting from Clifford's notation to mine:

$$a = 2i_1 + 3i_3$$
$$b = i_1 - i_2 + 2i_3$$

Under Clifford's product, their juxtaposed multiplication, ab, is $(2i_1 + 3i_3)(i_1 - i_2 + 2i_3)$, which any well-schooled teenager could calculate:

$$2i_1i_1 - 2i_1i_2 + 4i_1i_3 +$$
$$3i_3i_1 - 3i_3i_2 + 6i_3i_3 \ .$$

Clifford asserts that i_1i_1 and i_3i_3 each become 1, leaving 2 + 6. Clifford also asserts, for the remaining terms expressed in cyclic indices, that $4i_1i_3 + 3i_3i_1$ becomes $-i_3i_1$, and $-3i_3i_2$ becomes $3i_2i_3$, which leaves this sum: $2i_1i_2 + 3i_2i_3 - i_3i_1$. When all of this is expressed as a basis-composition, it becomes $8 + (2i_1 \blacktriangleleft i_2 + 3i_2 \blacktriangleleft i_3 - i_3 \blacktriangleleft i_1)$, which is simply a•b + a◄b, as you may check.

Now let us reverse this tactic and proceed from composite a•b + a◄b to juxtaposed ab. The retraction-plus-extension is composed of basis terms each having this form: $ab\ i_a$•i_b + $ab\ i_a$◄i_b. When $_a$ and $_b$ are different, these terms become simply $ab\ i_ai_b$ because their retraction vanishes. When $_a$ and $_b$ are the same, both $_a$ say, the terms $aa'\ i_a$•i_a + $aa'\ i_a$◄i_a become simply aa' because their extension vanishes.

Then, when all of these individual products are added together, their sum becomes ab. This ploy truly is a stroke of genius because it combines extension and retraction into a slightly enhanced *bottom-up* arithmetic.

Articulateness of extension-retraction

Combining two binary products to make a new one is

seldom a fertile idea. For example, combining addition and multiplication would almost never be useful. True, they are often used together, but only rarely with the same arguments. If that were ever needed, they could easily be combined separately.

Especially, combining two products that generate *different kinds of things*, like extension and retraction do, seems bizarre. Grassmann thought so, which is why he didn't do it until he had seen it done, just before he died.

Actually, what he had seen done was just the opposite: two products *separated* from one product, namely Hamilton's quaternion product. This *one* product had unexpectedly generated *two* kinds of things, a vector and a scalar.

The scalar startled Hamilton on the day of his discovery; and he initially thought it might somehow represent time. A half century later Hamilton's scalar entwined with his vector dismayed Heaviside and especially Gibbs, who simply hacked them apart:

> In regard to the product of vectors, I saw that there were two important functions (or products) called the vector part & the scalar part of the product, but the union of the two to form what was called the (whole) product did not advance the theory as an instrument of geom. investigation. [*A History of Vector Analysis*, p152]

So successful was Gibbs in separating these products that few users of his vector algebra would now ever try to combine them into a cross-plus-dot product, ×+•. ¿How could that ever work?—they generate different kinds of things! Well, it worked fine beginning in October 1843.

Clifford was not *explicitly* trying to combine two different products either, despite that consequence. He was aiming to generate a *single product* from which *different n-*compositions could be *separated "using selective symbols V_0, V_1...V_n analogous to Hamilton's S and V"*. He was seeking a *separable union*, the reverse of Gibbs.

For example, his union of a and b is ab, a single composite entity in his mind. Separation of the scalar composition 8 from it, which he would denote as as V_0.ab, is our retraction a•b, a distinct symbolism that likely would have seemed needless to him.

Separation of the the bivector composition $(2i_1i_2 + 3i_2i_3 - i_3i_1)$ from it, denoted as V_2.ab, is our extension a◂b, another apparently needless symbolism. His subscript on *V* tells the *number of basis vectors* in each term, his bottom-up way of establishing dimension.

(For dimensional separation, Hestenes introduced an enveloping *grade operator* \ *grade filter* \ *grade projection* (depending on author): ‹uv›$_0$ and ‹uv›$_2$ that completely displaced Clifford's \ Hamilton's symbolism.)

To define his unified separable product, Clifford began with the scalar product of Hamilton, for which multiplication of identical basis vectors generated the unit scalar. Then he proceeded to "*extend this assumption*" onto Grassmann's neg-commuting basis vectors.

In other words, he merely *replaced* Grassmann-like $i_si_s = 0$ with Hamiton-like $i_si_s = \pm 1$—his intent was to make Grassmann's multi-displacement algebra inform like Hamilton's three-vector algebra did. He was motivated by raw syntactic considerations in much the same way as Hamilton had been a half century before.

Altho he never claimed, nor intended, to be adding Grassmann's inner product to his outer product, as Grassmann eventually did; nonetheless that is precisely the *implicit* effect of his simple 1-for-0 replacement. So whether you focus on its whole or on its parts, his product is most meaningfully referred to as *extension-retraction*.

That gives us a choice of how to proceed: ¿Should we consider extension and retraction individually fundamental, with extension-retraction a derived product? Or should we consider unified extension-retraction fundamental, with extension and retraction derived from it?

The first path is geometrically enlightening; the second is algebraically elegant. Both paths are fruitful, but since we have already traveled quite a distance along the first one, let us restart and consider Clifford's unified product fundamental, at least provisionally. Its bottom-up ploy is most efficacious computationally, and also most transparent about its versatile algebraic properties.

Here they are: Unlike retraction, extension-retraction has an identity. Unlike in extension, every free element has an inverse under extension-retraction. Unlike retraction, extension-retraction is associative. However, like both extension and retraction, extension-retraction respects summary.

Consequently, this new product regains much of the *essence* of the field axioms for multiplication, except that it does not always commute, and it does not always give a mixed number an inverse.

However, it does not commute *in a very careful way* that allows extension and retraction to be extracted as odd and even parts, as we shall now see. Moreover, when it can't give a mixed number an inverse, that is only because *that number makes rich dimensional distinctions* that cannot be eliminated. In short, its *algebraic* failure to conform to the field axioms owes to its *geometric* success at informing.

So, combining extension and retraction is a fertile idea indeed, even tho they generate different kinds of things—or rather *because* they do.

Identity

One of the lovely properties of Clifford's bottom-up approach is that it considers scalars to be honorary *nullary* basis elements that remain aloof from his two rules. This is clear from his *selective symbol V_0*, which explicitly considers a scalar to be a valid *term* in a basis composition. When that composition subsequently participates in extension-retraction, its scalar thence become a valid *factor* too.

So let us extend-retract the scalar factor 1 with a free mixed number $\underline{M}$. That scalar simply multiplies each basis term in $\underline{M}$ by 1, leaving $\underline{M}$ unchanged. So 1 is the identity for extension-retraction. This seems almost too obvious to merit comment, but it is a *simultaneous* engagement of a *nullary* element with both extension and retraction.

If we could somehow *individually* engage those products with a *generic* element, then we could specialize that element to a nullary one. That would tell us how extension and retraction *individually* engage with a scalar, something that is not at all obvious from Clifford's description.

The usual way of doing that is to exploit complementary commuting properties like so:

$$\ell_{odd} \bullet v = (\ell_{odd}\, v + v\, \ell_{odd})/2$$
$$\ell_{odd} \blacktriangleleft v = (\ell_{odd}\, v - v\, \ell_{odd})/2$$

The easiest way to understand these equations is to specialize ℓ_{odd} to the most familiar odd free element you know, namely a free vector. Then recognize that all odd elements *commute* the same way under retraction with v, and *neg-commute* the same way under extension.

Strangely, that *commuting* is here defined by *addition*; and that *neg-commuting* is defined by *subtraction*. These equations simply subtract out the extension and duplicate the retraction; or subtract out the retraction and duplicate the extension; respectively, for which division by 2 is needed to un-duplicate.

(This is a classic contrivance often used to dismantle a function into its odd and even parts. The most well-known example is the dismantling of e^x into *cosh x* plus *sinh x*. Give that a try.)

Of course retraction and extension of an *even* element, ℓ_{even}, with a vector have the opposite commuting properties. To illustrate their *form*, let us engage them with the most familiar even element you know, the honorary one, namely a scalar s, even dimension {0}:

$$s \bullet v = (s\, v - v\, s)/2$$
$$s \blacktriangleleft v = (s\, v + v\, s)/2$$

Here, generic element ℓ_{even} has been specialized to nullary element s. (Notice the opposite signs compared to the odd equations.) All other even elements have the same commuting properties; but their specialization here tells you exactly how retraction and extension engage with a scalar, according to

Clifford:

$$s \bullet v = 0$$
$$s \blacktriangleleft v = s\, v$$

These results had previously been motivated *geometrically*, and now Clifford is affirming them *algebraically*. They enable a vacuous generalization of Grassmann's central product:

$$ab = a\bullet b + a\blacktriangleleft b$$

This is vacuous because the retraction with scalar a vanishes. However, when that scalar is 1 it has the virtue of demonstrating that retraction has no identity, tho extension does; so extension-retraction's identity derives from extension, not retraction.

The just seen odd–even, top-down definitions of extension and retraction are conceptually enlightening, and sometimes analytically useful, but they have two serious flaws.

First and most obvious, they are computational kludges: to generate a retraction, or an extension, Clifford's product must be engaged twice, once forward, once backward. Then these results must be half-canceled, half-duplicated, which nullifies half the work that had been done. Finally, the duplicated work must then be un-duplicated.

This is something that might have delighted Rube Goldberg after he had gotten a frightened rabbit to trip on a valve that turned on a lawn sprinkler that … that startled a donkey that pulled on a rope that opened a garage door.

Second and most evident, but maybe not at all obvious why, these definitions always engage an element with a *vector*, rather than another *element*. The reason is fairly simple: extension and retraction are iterated *vector* operations, for which each iteration produces something with *different dimensional parity*.

Consequently, extension or retraction of an element with another element requires *alternating* between even and odd equations *one vector at a time*—each operation must constantly switch out an even horse for an odd one, or vise-versa.

Computing extension or retraction of entire elements in that way, I hope it is clear, would be an *extreme* computational kludge: *most* of the previous work would be meticulously duplicated and then carefully thrown away.

Fortunately, there is slight discipline on Clifford's product that enables extension or retraction to be done all at once, quickly, on entire elements. It requires two interrelated keys, *basis tidiness*, coming up, and ...

Inverses

A multiplicative inverse of a free mixed number $\underline{M}$ is a number that multiplies it to the identity, called its *reciprocal*, and denoted as $/\underline{M}$. Reciprocals are rarities in geometric algebra because $\underline{M}$'s various dimensions typically prevent it from multiplying down to the dimension of the identity.

Fortunately, the basis elements that compose $\underline{M}$ can always be individually multiplied down to the scalar identity, even tho they generally have non-scalar dimensions themselves. Clifford's first equation provides the key: $i_s i_s = 1$.

This is his ***dimension-decrementing mechanism***, a fact that is seldom mentioned. It should be emphasized in ***bold italics***. (His *dimension-incrementing mechanism* is his second equation, which prevents an accreted product of distinct basis units from coalescing. This is seldom mentioned either.)

Before we can turn Clifford's key, we need notation leading to basis tidiness. A basis element is a scaled product of basis units, $-5i_2i_3i_1$ for example. Let us call such an element a *base*, and denote it generically by ellipsis: ... , or --- for a distinct base, or ,,, for a still different one, and so on.

At bottom, a mixed number is nothing more than a sum of various bases of various lengths, so it could be symbolized as ... + --- + ,,, for a three-sum; or as ... for a one-sum, and so on. A base such as ... may itself contain only a single basis vector, or even none at all if it is a scalar.

Such pithy notation may puzzle you now, but it shall shortly be enhanced with appendages that considerably simplify the unified axioms for the full geometric algebra.

To arrive there, the notation prepares us to find the reciprocal of --- = $-5i_2i_3i_1$. The reciprocal of its basis-unit part *** = $i_2i_3i_1$ is trivial: ↩*** which equals $i_1i_3i_2$. Clifford simply juxtaposes these, ***↩*** meaning $i_2i_3i_1i_1i_3i_2$. This all collapses to 1 by iterating $i_si_s = 1$ from the middle out.

Knowing that, the obvious reciprocal of the original base ---, namely $-5i_2i_3i_1$, is this somewhat intricate expression: ↩--- /---↩---, which equals ↩---/−5. This becomes $-1/5\ i_1i_3i_2$ when expressed in basis form.

The intricacy of this reciprocal expression can be avoided with a double-negative, dividend-divisor simplification that dispenses with reversals: ---/------. (Notice that Clifford makes the divisor a ±scalar because all its basis vectors are repeated.) Clearly, when the original base --- multiplies this, it

264

becomes ------/------, which is 1 under Clifford's product.

The geometric-algebra expression of this for a generic free element $\underline{e}$ is $\underline{e}/\underline{e}^2$, an expression already seen in this form: $\underline{e}/\underline{e}{\bullet}\underline{e}$. Extension does not have retraction's ability to multiply down to a scalar like this, so extension-retraction's inverses derive from retraction, not extension.

Associativity

Having just seen how Clifford simply juxtaposes two bases to multiply them, you might think that associativity is already implicit in his bottom-up approach because clearly …

$$(...\text{---}) \,,,, \quad = \quad ...\text{---}\,,,, \quad = \quad ... (\text{---}\,,,,)$$

This equation would automatically be true if it contained no repeated basis vectors. In that case ...---,,, would be a *pure extension* of these three bases.

However, if the expression does contain repeated vectors, it would vanish as an extension, a tidiness that would enforce extension's associative rule. But it would not vanish as an extension-retraction, an untidiness that requires checking the associative rule for that product.

We must check how it uses Cifford's second equation, $i_r i_s = -i_s i_r$, to walk duplicate pairs of vectors next to each other; and then uses his first equation, $i_s^2 = 1$, to eliminate them.

Pairwise, such duplicate-elimination is unambiguous regardless of the order in which it is performed: the same number of sign changes is required to walk a left duplicate to a right one, as to go the other way, or even to meet in the middle. The only potential ambiguity arises when pairwise canceling eventually descends to *three* remaining duplicates.

For example, consider this scenario:

$$.i_s.. \ --i_s- \ ,,,i_s$$

Grouping this left-to-right, $(.i_s.. \ --i_s-) \ ,,,i_s$, descends to this expression: $\pm...---,,,i_s$. Grouping it right-to-left, $.i_s.. \ (--i_s-,,,i_s)$, descends to this: $\pm.i_s..---,,,$. The $\pm$ signs indicate the overall sign change induced by neg-commuting one duplicate over to the other. ¿Do these different results have the same sign?

Yes they do. This is easier to rassle with in ungrouped form: $--i_s----i_s--i_s---$. The two ways of grouping, in essence, amount to either walking the left duplicate to the middle, or the right one to the middle, where they annihilate each other.

The first scenario leaves the right duplicate as residue; the second scenario leaves the left duplicate. These scenarios have opposite signs only if one journey had odd sign changes, and the other had even ones. In that case, walking the left duplicate to where the right would have been, or vice versa, has odd + even sign changes, which properly recovers their opposite signs. All other cases have the same signs.

Hence, extension-retraction is associative, even tho retraction alone is not.

The reason retraction is not associative is that subtraction of dimension is not. For example, consider retraction of a bivector, a vector, and a trivector: $(B•v)•T$. If this doesn't vanish the first retraction produces dimension $\{2-1\} = \{1\}$; the second produces dimension $\{3-1\} = \{2\}$.

Conversely, for $B•(v•T)$ the first retraction produces dimension $\{3-1\} = \{2\}$; the second produces dimension $\{2-2\} = \{0\}$. Clearly these results are not the same. To say this geometrically, retraction loses dimensional information, and the sequence of its loss makes a difference.

Extension, on the other hand, *can be* associative because addition of dimension is: $(2+1)+3 = 2+(1+3)$. However, validating such *algebraic potentiality* requires corroborative *geometric meaning*, already provided in the *Extending* chapter. It showed exactly how extension gains dimensional information geometrically, for which the sequence of its gain makes no difference.

When extension-and-retraction are used together, the dimensional information that extension gains compensates for the information that retraction loses; so extension-retraction's associativity derives from extension, not retraction.

Consequently, the *algebraic assertion* of extension-retraction's associativity, as devised by Clifford, would not be valid if the *geometric fact* of extension's associativity were not valid. Semantics engenders syntax, you know.

Joy of bivectors

Now that you have seen how articulate combined extension-retraction can be, compared to co-articulated extension and retraction, your best strategy is to begin articulating it. For that purpose free bivectors are the absolute best elements to begin with. Nearly all books on the free geometric eagerly extend-retract them in multiple ways.

I have in front of me right now maybe a dozen such books. The three most-popular ones, *New Foundations for Classical Mechanics* by Hestenes, *Geometric Algebra for Physicists* by Doran and Lasenby, *Geometric Algebra for Computer Science* by Dorst, Fontijne and Mann, all have large sections that I call

Ode to the Bivector!

They all begin by showing that it squares to -1: $(i_1 i_2)^2 = i_1 i_2 i_1 i_2 = -i_2 i_1 i_1 i_2 = -i_2 i_2 = -1$. So a bivector functions as an enhanced *imaginary number* newly endowed with rich geometric meaning.

Then these books proceed to show that one vector times another generates a scalar plus a bivector, $a+bI$, which functions as an enhanced *complex number* also endowed with geometric meaning, making it better called a *spinor* or a *rotor* —anything but a "*complex number*".

The reason for such nomenclature is that, whereas the scalar merely scales a vector, the bivector rotates it by a quarter turn. For example $i_1\, i_1 i_2 = i_2$. Gleeful about that, these books then visually demonstrate that simultaneously scaling a vector and quarter-turning it can generate any kind of rotation, or rotation-dilation …

… in any orientation!

… in any dimension!

… and the ploy generalizes to other elements!

Woo Hoo!

I completely share their evident enthusiasm; and it was the expressive versatility of bivectors that initially sold me on the free geometric sub-algebra so long ago. (I had no clue then that it was free, or a sub-algebra, a common naivety even now.)

However, if you are not a *Classical Mechanic*, or a *Physicist*, or a *Computer Scientist*, the jargon and presumptions in these books may frustrate you, or dismay

268

you, or make you exclaim ¿*What on earth does this have to do with life, or the real world*? as happened to my sister.

That is the usual response to mathematics. It could be avoided if high school freshmen were gently introduced to the free bivector—I can't imagine anyone not liking mathematics after seeing such a beautiful number. But that would require elementary-school students first being gently introduced to addition of points, beautiful numbers in their own primitive way; and middle-school students being gently introduced to their extension.

¿Might the prospect of such gentle introductions excite a writerly impulse in you? The field is currently wide open—get in on the ground floor! I expect high demand within just a couple of centuries. Ha—sadly.

(Insider tip: Begin with a booklet, *Joy of Bivectors*. Then write precursor booklets that would let children understand it: *Joy of Point Addition. Joy of Point Subtraction. Joy of Doing Point Multiplication. Joy of Undoing Point Multiplication.* ...)

Inarticulateness of extension-retraction

Said simply, whereas the extension part of extension-retraction is always geometrically meaningful for any elements, free or bound; and whereas the retraction part is always meaningful for free elements, it is usually meaningless for bound elements.

But not inexorably, so let us begin meaningfully. Grassmann pioneered a fundamental ambiguity that seemed to allow free or bound elements to be meaningfully extended or retracted: he simply established an algebra of abstract "*units*" that could be interpreted either as points or free vectors, as you wished, and even intermixed if you liked.

Clifford subsequently promoted that innovation with his "*n* units i_1 i_2 ... i_n", which he initially illustrated with a product of points; and then switched to a product of free vectors, as tho his enhancement would work equally well on either. However, most people thereafter began to deploy his algebra on free-vectors exclusively, and for excellent reasons, which will become obvious shortly.

In contrast, the Grassmann-Algebra community persisted with vector-or-point units, as you wish; a cultural difference that continues to this day. This despite Grassmann finally realizing that retraction only works on free elements, as he indicated in his second book, CHAPTER 5, *Applications to Geometry*:

> 330. DEFINITION. For inner multiplication I always take as the original units in space three mutually orthogonal and equally long displacements (e_1, e_2, e_3), and in the plane two of them (e_1 and e_2), and in particular I take the length of these displacements as the unit of length and [$e_1e_2e_3$], and in the plane [e_1e_2], as units of volume and surface area.$_{.p176}$

This enlightenment was a long time dawning. Grassmann initially presented his inner product in an essay, *Geometric Analysis*, published in 1847 (this was Hamilton's first exposure to the inner product). It was a hastily written submission for a prize competition to see if anyone had made progress toward Leibniz's idea of a *Universal Algebra of Everything*. Grassmann believed that is *exactly* what he had progressed toward, and largely achieved; and he said so with barely restrained enthusiasm.

He launched his inner product with vectors projected onto

vectors, and then onto bivectors and so on up. The Pythagorean Theorem allowed him to develop a nascent inner algebra of vectors in a way complementary to his mature outer algebra of them.

After sketching that algebra, he began section §14. **TRANSITION FROM DISPLACEMENTS TO POINTS**[9], where things began to get curious—here is editorial comment 9: *"The theorems and definitions presented here are, as they stand, scarcely understandable."* August Mobius concurred in a commentary requested by the prize committee:

> The study of the preceding article, and particularly of its last part (from 14 on) may, despite the unmistakeable endeavor of its author for clarity, nevertheless be associated with some difficulties, which result because the author sought to base his new analysis on a method that lies rather far from the currently standard procedures of mathematical investigations, and because in analogy with arithmetic operations he treats objects as magnitudes that are not magnitudes as such, and of which one can in part form no representation.[p385]

Elsewhere, Mobius was more specific: *"No one will be able to imagine something intuitive by the inner product of a point with a line* [displacement]*, which is called a planar magnitude here, or by the inner product of two points, a spherical magnitude."*[p413]

Despite these misgivings, Mobius assented to award of the prize to Grassmann, possibly for his raw creativity, and possibly because his submission was the only one.

It is a tribute to Grassmann's innovative genius that *he*, personally, was able to imagine those products, vividly in fact; and it is a testament to his geometric sensibility and mental flexibility that he eventually abandoned them.

But he never explicitly proscribed them. Instead he

persisted with his abstract point-or-free-vector units, your choice, until he died; and relied on his readers to not formulate geometric nonsense. The merits of this strategy shall be entertained in the next and last chapter, *Geometric Algebraic Clarity*.

The geometric nonsense that bound retraction alone can generate has already been explored in the *Retracting* chapter; so let us now see what kind of nonsense bound extension-retraction can generate.

For that purpose we need at least one point amongst Clifford's units, but the Grassmannian ploy of simply reinterpreting a vector as a point would mask the nonsense. For good nonsense-detection we need a new point unit that is *formally distinct* from the old free-vector units.

So let us explicitly append a point to Clifford's free-vector basis: i_1, i_2 ... i_n, $\boldsymbol{i_o}$. (I stashed it on the right in anticipation of free-part extraction, which shall be effectively computed via *right retraction-by* the origin)

The new bound unit moves Clifford's algebra beyond a purely *free* one to a **bound** *free-as-possible* one. It has subscript o to indicate that it is intended as the *o*rigin. Its symbol *i* engages it with Clifford's two rules.

When engaging those rules, we of course notice $\boldsymbol{i_o}$'s informal **point** typeface, but the rules themselves don't. To get the algebra to formally notice that $\boldsymbol{i_o}$ is a point requires semantic axioms declaring that, whereas the old i_s units have dimension $\{1\}$, this new unit has dimension $\{\mathbf{1}\}$. That shall give the fruits of Clifford's rules crucial free-vs-bound distinctions they currently lack.

272

Clifford's first rule declares that $i_o^2 = 1$. This seems eminently meaningful: it is a self-retraction that generates the origin's square weight, like the self-retraction of a free vector generates its square separation.

However, the origin is the *only point* whose self-retraction does that. Consider $p = i_o + 2i_1$, for example, whose square is $i_o^2 + i_o 2i_1 + 2i_1 i_o + (2i_1)^2$. This equals its square weight 1 plus its square separation from the origin, 4. All non-origin points chuck in their separation from the origin like that.

Doing so is not yet contradictory, and not quite nonsense either; but it makes point retraction bizarrely context-dependent in a crippling way—its result depends on the arbitrary placement of the origin.

Clifford's second rule declares that $i_s i_o = -i_o i_s$. This is meaningful because it is a pure extension, $i_s \triangleleft i_o$, like every product of distinct Cliffordian units. It represents a bound vector whose length is the separation of its free part, i_s, namely 1. Its extension \ Cliffordian-product in the opposite order negates it. Point extension is definitely meaningful.

Even point retraction seems meaningful when it vanishes, as does $i_s \bullet i_o = (i_s i_o + i_o i_s)/2$. This seems to be a vacuous kind of perpendicularity similar to the vacuous perpendicularity of a scalar—neither has any projection onto free vector i_s.

So let us test vanishing point-retraction and non-vanishing point-extension on a generic free vector v times a point p—let us examine how Clifford computes their product vp.

This separates into $\langle vp \rangle_0$ and $\langle vp \rangle_2$. We would naturally expect the first to be v$\bullet p$, about which we have just acquired a good idea; and v$\triangleleft p$, about which we are pretty darn sure. Let us see if Clifford concurs.

To extract the 2-composition $\langle vp \rangle_2$ we first need to express v and p in terms of Cliffordian units. So let us say that v is $i_1 + 3i_2$, and we already know that p is $i_o + 2i_1$. That last, $2i_1$, is i_o's *translator*, which is going to be play a staring role, so let's label it t_r. It is the last vector in this expression:

$$vp \;=\; (i_1 + 3i_2)(\, i_o + 2i_1)$$

The 2-composition is simply the sum that contains terms with 2 distinct units: $i_1 i_o + 3i_2 i_o + 6i_2 i_1$. The first two terms constitute a vector bound from the origin. The last term is the translator to its confining space.

This is indeed $v \blacktriangleleft p$, expressed more informatively as $v \blacktriangleleft (i_o + t_r) = v \blacktriangleleft i_o + v \blacktriangleleft t_r$. The origin-bound vector is $v \blacktriangleleft i_o$; its translator is free bivector $v \blacktriangleleft t_r$. One end of that bivector cancels $v \blacktriangleleft i_o$, leaving the other end as residue. This is just $v \blacktriangleleft i_o$ shifted by a distance of t_r.

With that success, let us shift our attention to the 0-composition $\langle vp \rangle_0$. The good idea we have about this is that it is merely their retraction, which should vanish because a free vector and a point have no projection onto each other. Let us see if we are right:

The 0-composition is simply the sum that contains terms with repeated units. There is just one such term, $2i_1 i_1 = 2$; but that is one too many—we were expecting none at all. Nonetheless, this truly is the retraction $v \bullet p$ as we have been computing it for several chapters:

$$v \bullet p \;=\; v \bullet (i_o + t_r) \;=\; v \bullet i_o + v \bullet t_r \;=\; 0 + 2$$

Now the problem is clear: altho v's retraction with the

origin does properly vanish because i_o's retraction with every basis vector in v vanishes, by Cliffordian fiat; nonetheless *v's retraction with the point's translator* t_r *does not vanish.*

This is no longer just bizarrely context-dependent, it has become **geometric nonsense**. The *un-extension* of a point and a free vector *must* vanish because it can produce nothing by which a free vector could ever extend to a point; or by which a point could ever extend to a free vector. *Nothing.* 0.

And yet we have just seen v's *algebraic* retraction with a non-origin point not only producing *Something*, but something that depends on the arbitrary placement of the origin. This not just nonsense, it is *ambiguous* nonsense —*random* nonsense.

This kind of geometric nonsense is almost always generated when bound things are algebraically retracted. It is produced in our new bound Cliffordian algebra via translators that retract out spurious and nonsensical information about the arbitrary placement of the origin.

Algebraic **extension** *of bound things*, on the other hand, always generates good geometric sense. The translators in each argument extend out the translator of the result. Such a translator is indifferent to the placement of the origin like so: when the algebra is expressed in a different origin, its generating translators change to enforce a unchanged location for its result.

Algebraic *retraction of free things* also generates good geometric sense. Un-extension must presume the juxtaposition that extension had engendered when it swept one argument from the other. That presumption is effectively enforced in the free algebra by freedom and lack of translators. Their presence in a Cliffordian bound basis violates the presumption, which then generates retractive nonsense in that algebra.

This nonsense can easily generate *geometric* contradiction, as seen in the *Retracting* chapter; but it does not generate *algebraic* contradiction like division by zero does. *Semantic* contradiction: yes; *syntactic* contradiction: no.

That is fortunate because it allows you to use extension-retraction on bound things without fear of invalidating its extension part. Its versatile algebraic properties, especially associativity, make it a valuable tool even on them—you can use it as an efficacious way of performing extension; and then either ignore the retraction part, or filter it out dimensionally.

(Sometimes the retraction part is not nonsense—origin-bound elements often retract meaningfully. However, their bondage to the origin makes them context-dependent and relatively uninformative—pre-free-part extraction is almost always a better tactic.)

Of course if you don't need extension-retraction's versatility, you would better use pure extension, even in the free sub-algebra. And pure retraction there too—sometimes those operations are more adept in the free algebra than extension-retraction.

To see that, put yourself in Clifford's frame of mind, presuming he had lived long enough to develop the fruits of his new product; and that he had paid attention to how it came to be used. He would have observed that people typically deploy it primarily on *vectors*, and in two distinct ways:

First, it is used as a pithy tacit way of *extending* a vector that is perpendicular to an element; or else as a tacit way of *retracting* a vector that is parallel to an element. Such retraction is most frequently nothing more than Cliffordian-squaring an element, especially a vector.

Second, when his product is used with a vector that is skewed non-perpendicular and non-parallel to an element (which is often a bivector), it is done to exploit the full information it provides about their relative perpendicularity and parallelity; and also to exploit the versatile associative property, both often needed to perform rotations or establish multiplicative inverses.

His product is much less often used on skewed non-vector elements, especially large ones. And when it is used, it is very seldom left unfiltered—the result simply provides more information than needed, often *much* more; and that information is generally so intricate as to be incomprehensible to a person.

So a single dimension is typically extracted from such a product; and it is usually a boundary one: either the dimension for the pure extension of those elements, or their pure retraction.

For example, the extension of an *m*-vector $_m\underline{e}$ and an *n*-vector $_n\underline{e}$ would be done like this: $\langle _m\underline{e}\,_n\underline{e}\rangle_{m+n}$. This is far more efficient than computing their extension using the odd-even, top-down, one-vector-at-a-time equations previously developed; but it still does a lot more work than needed, and then throws most of it away.

By contrast, pure extension and pure retraction, when made efficient as sub-operations of Clifford's extension-retraction, would do only the minimal amount of work necessary, and never throw anything away. Specifying such sub-operations is a crucial task in …

Unifying the full bound-and-free algebra

As a prelude to tri-unifying extension, retraction and extension-retraction for expressiveness and efficiency, we need to unify terminology in two ways: a more-evocative name for the last operation, and a more-informative notation for bases.

Altho "*extension-retraction*" is the most meaningful way of expressing Clifford's product *iteratively*, it is long-winded; and it is somewhat misleading about the intertwined intricacy generated by that product.

To begin appreciating its intricacy, consider Clifford-multiplying a free m-vector $_m\underline{e}$ with an n-vector $_n\underline{e}$, supposing that m is less than n. Each base in $_m\underline{e}$ multiplies each base in $_n\underline{e}$; and the duplicated units in each such product can range from none at all to complete duplication.

For no duplicated units, Clifford's product generates a new base with dimension $m+n$. For one duplicated unit, the product generates dimension $m+n-2$. For two duplicated units, $m+n-4$. And so on down to a new base with dimension $n-m$ for complete duplication.

The sum of these new, variously dimensioned bases is the intricate basis composition for that product. Clifford's paper rassled primarily with its numeric intricacy. About that, there are really only two readily understandable dimensions: the $m+n$ dimension, which is pure extension; and the $n-m$ dimension, which is pure retraction. And even those are impossible to visualize beyond dimension {*3*}.

278

So it hardly seems appropriate to call all of these various dimensions "*extension-retraction*" when only two of them are. ¿Maybe *dimensioned extraction* would be better? Or plain **extraction**? That term, as a bonus, is a contraction of **extension-retraction**. What a coincidence.

So that is what I shall call it hereafter. I get to do that in **my** book—you get to call it whatever you want in **your** book.

Conceptually, extraction is always performed by walking duplicate Cliffordian units next to each other where they suddenly become 1, and vanish. There are different tactics for walking which leads to ...

Basis tidiness, part A

Each base in a basis composition is a scaled extension of basis units, usually pure free vectors i_s, but occasionally including the origin i_o (about whose tidiness this section has more to say at the end).

Each base is of course expressed without duplicates—they would annihilate that base under Grassmann's outer product; they would just disappear under Clifford's product; and their presence would preclude using the unit-count in a base as the specifier of its dimension.

This is an *implicit* discipline, almost never deemed worthy of explicit mention. The advantage of making it explicit is that it forces focus on just who is enforcing the discipline.

In Clifford's algebra the enforcer is extraction; and it must decide on how to walk duplicate units together to eliminate them. ¿Should it walk them together in the order they fell to earth, and eliminate each pair individually? Or should it first walk them all together, and then eliminate them all at once? (Final answer in part B: none of the above.)

Preliminary hint for part A: Extension, retraction, extraction—to eliminate duplicates—all require them to first get juxtaposed somehow; but then those different products each eliminate them in different ways.

So if we first walk all duplicates together in a canceling configuration, then each operation can take that result and eliminate them in its own peculiar way. The walking operation is specified by extraction's neg-commuting rule, $i_r i_s$ = $-i_s i_r$; but extension and retraction both need it to do their walking too.

So let us extract that rule from extraction and install it in a *pre-product percolation ploy*, to wit: Whenever we juxtapose two bases together in preparation for their extension or retraction or extraction, their duplicates percolate together like so:

$$... \text{,,,}\quad \text{becomes} \quad .. \text{»}i * i\text{«} \text{,,}$$

On the left you see two bases juxtaposed in preparation for percolation. On the right you see their duplicate units percolated together in reverse order: »i = $↶i$«. The sign of percolation in each base is absorbed in its scalar—that scalar ensures that each product gets its sign right.

The asterisk $*$ is the *pre-product* symbol. It will be replaced, during the subsequent duplicate-elimination phase— the *basis-tidiness* phase, by ◂ for extension, • for retraction, and simple juxtaposition for extraction. Here is an example percolation with duplicates emboldened:

$$4i_7i_2i_5 \; 3i_1i_9i_5i_4i_7 \text{ becomes } -4 \; i_2 \text{ ›}\mathbf{i_7i_5} * \mathbf{i_5i_7}\text{‹ } i_1i_9i_4 \; 3$$

To help the basis-tidiness phase do its job properly, percolation will produce .. * „ if there were no repeated-units.

If there were, percolation with produce .»›i * i‹‹ „ if the left unrepeated-units were absent, .. »›i * i‹‹, if the right unrepeated-units were absent, or .»›i * i‹‹, if they both were. These are mere refinements of .. »›i * i‹‹ „

The single dot or single comma attached to a repeated-units indicates a base scalar that has been transferred to it by absence of its unrepeated-units.

That is why the scalar in the right base was placed on its right after percolation—that tactic automatically transfers that scalar to its repeated-units whenever its unrepeated-units are absent. This is a crucial discipline needed to unify scalar multiplication with the other products.

When a base is a pure scalar, there is no possibility of it having any repeated units with the other base, so percolation will generate the .. * „ pattern refined to more informative . * „ for a left scalar, .. * , for a right scalar, or . * , for left-right scalars.

Left-vs-right distinction is overkill for scalars, but it keeps the extension-retraction-extraction rules uniform and simple; and it sidesteps the need for separate scalar multiplication rules, as we shall now see. It functions partially as an implicit commuting rule for scalars.

Extension basics

To begin appreciating the expressiveness of this notation, let us begin using it on the most basic operation, extension. That operation would be specified like so:

$$.. \underline{\text{»i}} \triangleleft \underline{\text{i«}} \,,, \quad \text{becomes} \quad 0$$
$$.. \triangleleft \,,, \quad \text{becomes} \quad ...,,$$

The pre-product symbol, *, has been replaced by the extension symbol ◄. Its first rule says that if *repeated*-units are *definitely present*, the extension vanishes. This is extension's *abort rule*, its recognition that these two bases are at least *partly parallel*. It is extension's extreme way of enforcing basis tidiness.

The *definitely-present* indicator is the <u>underline</u>, which shall be even more crucial for specifying retraction. The lack of an underline on the unrepeated-units .. and ,, in this rule means they may be absent: the left unrepeated-units .. might be absent, or the right ones ,, might be, or both.

(In such cases, the base scalars would automatically transfer to the repeated-units, a moot point here since this extension vanishes.)

The second rule says that if the two bases contain no repeated-units, their extension is simply their Cliffordian left-right concatenation. In that case the two bases were *completely perpendicular* under Clifford's expressive presumptions.

(These bases do not require a definitely-present underline because all rules presume *Something* is present on each side. In this case it is .. ◄ ,, or refinements of it: . ◄ ,, or .. ◄ , or . ◄ ,)

Now you might think that this is still not the most-efficient way to compute extension. You would be right because the percolation for *this* operation could have been aborted after finding just one duplicate.

Nevertheless, it is the most-efficient way to *specify* extension in a unified way with the other operations—it is a Cliffordian *bottom-up* enhancement of extension. How it gets engaged from the top down shall be examined as a prelude to the unified axioms. After they are specified, computational efficiency shall be rassled with.

One of the benefits of this specification is how transparently it exposes scalar extension with *anything*—not just with a vector—as nothing more than scalar multiplication.

To see that, suppose the left base were a pure scalar, 5 say. Percolation would have presented extension with 5 * ,, which engages a refinement of extension's second rule. Its left-right concatenation then produces 5,, meaning 5 times the unrepeated-units on the right.

Such concatenation is true even if the right base were a pure scalar too; so enhanced extension subsumes scalar multiplication, whether of non-scalars, or of scalars themselves. This is in perfect accord with extension's commuting properties: scalars, as even (nullary) "elements", commute with everything under extension.

You know that mixed-number *summary* also subsumes scalar *summary*, whether of non-scalars, or of scalars themselves. Therefor, since extension respects summary, *enhanced extension completely subsumes scalar multiplication*—there is no need for the separate scalar rules that I listed previously, rules that infest the exposition of vector spaces.

However there is some question about the meaning of scalar extension. For scalar ◄ vector extension the meaning is

that the vector is extending dimension {*0*} up to dimension
{*1*}. So by extension, scalar ◄ *n*-vector extension would mean
that the *n*-vector is extending dimension {*0*} up to dimension
{*n*} one vector at a time—not very puzzling.

Slight puzzlement begins with perpendicularity—the
second rule only engages when the units in the two bases are
completely perpendicular, meaning that a scalar is evidently
perpendicular to every element. For this to be valid, its
retraction with every element must vanish. We shall rassle
with that shortly.

More puzzlement ensues when we ponder scalar ◄ scalar
extension—scalars have no capability to increment
dimension, so how could they possibly extend? This clearly
works algebraically as pure scalar multiplication, but what is
its geometric meaning? The next section shall resolve that;
but somewhat cavalierly.

Extraction encompassment

Now we are ready to specify in an enhanced way the most
encompassing operation—*extraction*—Clifford's original
product:

$$.. \gg i \; i \ll \; ,, \quad \text{becomes} \quad ..,,$$

The pre-product symbol * has been replaced here by
Clifford's wonderful symbol, nothing at all—mere
juxtaposition. There is no chance of mistaking this for
percolation, which also uses juxtaposition—the presence of ≫i
i≪ indicates that these two bases are presumed to have already
been percolated together.

Notice how simple this rule is—nothing must be definitely
present (presuming *Something* is present on each side). If the

284

reversed repeated-units are present, they annihilate each other, which is extraction's maximally simple way of enforcing basis tidiness.

In such a case, the unrepeated-units may be absent on the left, producing .,, or on the right, producing .., or both, producing ., In these sub-cases, extraction is effectively pure retraction; and the final sub-case generates a pure scalar.

Conversely, if the repeated-units are absent, then this extraction is effectively pure extension. Which raises this question: ¿Does enhanced extraction also subsume scalar multiplication like enhanced extension does?

Yes it does, and the reason is ridiculously simple: Whenever either base in an extraction is a pure scalar, that extraction performs extension because percolation would not have presented it with any repeated-units. Instead it would have presented scalar refinements to the unrepeated-units, just as before. This induces scalar multiplication via left-right concatenation, as you saw.

Notice that in such a case extraction is not *deferring* to extension, any more than it ever defers to retraction; it is simply performing extension in its own encompassing way. Which somewhat reduces the puzzlement about scalar ◂ scalar extension:

When such extension is performed indirectly by a product that encompasses the others, as extraction does, it would naturally be expected to encompass scalar multiplication too —the puzzlement about a 3 ◂ 5 extension is completely absent in a 3 5 extraction, even tho these different notations perform identical computations. Admittedly, this is a cavalier *semantic* simplification overlaid on a *syntactic* one.

But it is a wonderful simplification because it allows pithy juxtaposed notation to be used for extraction and scalar multiplication indifferently; and it dispenses with the need for separate axioms for scalars.

Of course un-pithy extension notation could also be clumsily used for scalar multiplication. However, for compositional clarity that notation is best reserved as a fallback for when a scalar inadvertently arises in an extension; or in preparation for its undoing by retraction.

I personally regard even inadvertent extensions like $3 \triangleleft 5$ as an indication that I likely formulated geometric nonsense, and should better have used extraction than extension.

Retraction intricacies

Here is how retraction becomes specified in an enhanced Cliffordian bottom-up way:

$$\text{.. } \gg i \bullet i \ll \text{ ,, } \quad \text{becomes} \quad 0$$
$$\text{.}\gg i \bullet \underline{i} \ll \text{ ,, } \quad \text{becomes} \quad \text{.,,}$$
$$\text{.. } \gg i \bullet \underline{i} \ll \text{, } \quad \text{becomes} \quad \text{..,}$$
$$\text{.}\gg i \bullet \underline{i} \ll \text{, } \quad \text{becomes} \quad \text{.,}$$

The pre-product symbol * has been replaced here by Gibbs' dot •. Its first rule says that if *unrepeated*-units are definitely present in each base, the retraction vanishes. This is retraction's abort rule, its recognition that these bases are at least *partly perpendicular*. It is the first of retraction's several intricate ways of enforcing basis tidiness.

Its next three ways of enforcing tidiness have already been seen as implicit possibilities in extraction. They must be explicitly spelled out for retraction because there are three different ways that bases can be *not even partly perpendicular*, meaning *completely parallel*. (Extraction does not care about that one way or the other).

Completely parallel pays attention to relative size of the bases: it is always the little guy that is parallel to the big guy —*parallel-to* is like a *poorer-than* relation. It induces these cases for retraction: little•big, big•little, same•same.

(*Completely perpendicular* does not pay attention to size: *little-perpendicular-to-big* means *big-perpendicular-to-little* —*perpendicular-to* is like a *friend-of* relation. That is why extension required only one rule for that case.)

Here you finally see algebraic manifestation of the *little-guy-retractor* ploy that had been motivated geometrically in the *Retracting* chapter. You also see algebraic manifestation of that-chapter's iterative *one-vector-at-a-time* extension undoing.

To observe both manifestations clearly, look at a specific example of little•big retraction, denoted generically as .»i • i‹‹ ﮯ :

$$-4 \; ›i_3i_7i_5 \bullet i_5i_7i_3‹ \; i_1i_9i_4 \; 3$$

Here you see two base extensions that have already been percolated together. Retraction replaces the pre-product symbol * with •, which engages its rules. The first rule does not match, but the second does, as shown.

That rule precipitates out -12 ﮯ in one fell swoop, which equals $-12 \; i_1i_9i_4$. It shall be enlightening to instead do this iteratively via Cliffordian concatenation, one pair of units at a time:

$$-12 \; i_3i_7i_5i_5i_7i_3i_1i_9i_4$$

Beginning inside-out, the first iteration collapses i_5i_5 to 1. The next iteration collapses i_7i_7 to 1. And so on. This owes to Clifford's first rule, which is his way of *unextending-by-fiat*. This computation is simply a wonderfully pithy way of deploying the extension-undoing equation presented in the *Retracting* chapter, like so:

$$-12\ i_3 \bullet (i_7 \bullet (i_5 \bullet (i_5i_7i_3i_1i_9i_4)))$$

This clearly exposes the *little guy doing the retracting one vector at a time*, a perspective that is hidden in Clifford's condensed notation. If Clifford's inside-out unextension had begun right-of-middle, rather than left-of-middle, the extension-undoing equation would have instead iterated from right to left one vector at a time. The little guy would still be doing the retracting.

Now we are ready to rassle with retraction with a scalar. In this case percolation does not present retraction with repeated-units—they went absentee, leaving unrepeated-units definitely present, but in scalar-refined form. This engages the first rule, producing *Nothing*: retraction of anything with a scalar, even another scalar, vanishes. So scalars are vacuously perpendicular to everything, even themselves.

The meaning of that for scalar•vector retraction is that the vector is attempting to decrement below dimension $\{0\}$, but is being blocked. The meaning for scalar•n-vector retraction, is that the retraction is being blocked n times.

The meaning for scalar•scalar retraction seems to be even less than for scalar◄scalar extension; with virtually none of its algebraic usefulness.

Nonetheless, scalar•scalar retraction can arise in the algebra, and its formal disappearance gets it out of the way without generating syntactic contradiction. However, I consider it as meaningless as retraction with a bound element often is; so I reformulate my algebra to avoid it. That usually means using extraction rather than retraction.

There remains one more bottom-up enhancement to Clifford's algebra that is crucial for eloquence in the full algebra. It moves bound things into the free sub-algebra where they can engage retraction and extraction without generating nonsense:

Free-part extraction

You may have noticed that, aside from no duplicates in a base, the basis hasn't been all that tidy so far—the units in a base were displayed higgly-piggly in no apparent order.

This turns out to be unacceptable for efficiency, so *Basis tidiness, part B* will keep all units in a base well ordered, and then carefully *zip* them together in tailored ways. We shall now pre-exploit that tidiness to make free-part extraction efficient in a bottom-up way.

In a well-ordered base, whenever i_o is present it will be the last unit in that base (a convention explained shortly). Its presence indicates that base is bound thru the origin; so extracting its free part amounts to nothing more than removing its last unit.

This is a somewhat mystifying bottom-up ploy, in contrast with the transparent top-down development previously presented in this chapter. It depends on having already learned that every bound element is a free element extended from a point. Unextending by that point recovers the free element.

Once you understand those semantics, you can encode

them efficiently in syntax, similar to what Clifford did in 1876.

The other semantics you need to understand is that the free part of a free element vanishes because it *effectively* becomes a sum of exactly opposite free elements.

This depends on having already learned that every free element is a sum of *separate-but-otherwise-exactly-opposite* bound elements. You would have no clue about that just from looking at the bound enhancement to Clifford's *free-as-possible* basis—you would need to have started with a pure point basis to understand that.

But having understood all of that, you can now generate the following pithy syntactic rules for free-part extraction:

$$[]..\underline{i_o} \text{ becomes } ..$$
$$[].._{i_o} \text{ becomes } 0$$
$$[](\underline{M} + \underline{N}) = []\underline{M} + []\underline{N}$$

The first two rules are bottom-up, each defined on a base. The first rule says that if a base definitely ends with the origin, its free part is simply its non-origin part; and this includes a pure scaled origin, in which case its free part is its weight. This is a syntactic encoding of the semantic fact that a free part is effectively *right-retraction-by* a point.

The second rule says that if a base definitely does *not* end with the origin, its free part vanishes, as explained; and this includes a pure scalar. All of this causes the free part of any free element to vanish because every one of the bases that compose it will.

It also causes the free part of a bound element to pay attention only to its origin-bound part, and ignore its translator. And more than that, it also ignores the associated free non-translator part. Specifically, it ignores the translator of a bound-vector thrust, and also ignores the residual free-bivector twist.

The third rule is free-part-extraction's respect for summary. It functions as a connective top-down rule, like so: Reading from left to right, it recursively distributes the free-part operator down to the individual terms in a summary composition. When that operator finally descends to a base — as it eventually does for every base in the composition — it engages the bottom-up rules.

Each one of the other bottom-up rules need this kind of top-down rule to connect it to mixed numbers, or even to simple elements; so you may have wondered: ¿Where were those rules for extension, retraction, extraction? The answer is that they can all be specified in a single pre-product rule, yet another benefit of the *pre-product percolation ploy*. That rule is a star player in the …

Unified syntactic axioms

Coherence

$$\underline{M} + \underline{N} \;=\; \underline{N} + \underline{M}$$
$$(\underline{M} + \underline{N}) + \underline{P} \;=\; \underline{M} + (\underline{N} + \underline{P})$$
$$\underline{M} + 0 \;=\; \underline{M}$$
$$\underline{M} + (-\underline{M}) \;=\; 0$$

Percolation

$$\overset{\curvearrowleft}{}\!i_a \ldots i_n \;=\; i_n \ldots i_a$$
$$i_{\langle\langle} \;=\; \overset{\curvearrowleft}{}\!_{\rangle\rangle}i$$
$$1\ldots \;=\; \ldots$$

$$i_r i_s = -i_s i_r$$
... „ becomes .. ›› i * i‹‹ „
with refinements, as discussed

Respect for coherence
$$\underline{M} * (\underline{N} + \underline{P}) = \underline{M} * \underline{N} + \underline{M} * \underline{P}$$
pre-product * is specialized by ...

... extension
.. ›› $\underline{i}$ ‹ $\underline{i}$‹‹ „ becomes 0
.. ‹ „ becomes ...„

... retraction
.. ››i • i‹‹ „ becomes 0
„››i • i‹‹ „ becomes .„
.. ››i • i‹‹, becomes ..,
„››i • i‹‹, becomes .,

... extraction
.. ››i i‹‹ „ becomes ...„

Free-part extraction
$^{[]}.._{\underline{i_o}}$ becomes ..
$^{[]}.._{i_o}$ becomes 0
$$^{[]}(\underline{M} + \underline{N}) = {}^{[]}\underline{M} + {}^{[]}\underline{N}$$
unary respect for coherence

The hoary rules of addition have here been renamed the rules of *coherence*, terminology that is more accurate for mixed numbers.

Addition arose historically as a *coalescence* of many *same*-things to one; but in geometric algebra its main function is to *cohere* many *different*-things together. When such cohering

292

does encounter same-things, it often coalesces them, but only as a sub-function of cohering.

This idea is only just now beginning to leak outside the confines of mathematics. Even within mathematics its full import has still only been half appreciated. Clifford was the first to explicitly expose it a century and a half ago with his "*selective symbols V_0, V_1...V_n analogous to Hamilton's S and V*".

Hamilton's S and V were the first dramatic appearance of *explicitly* different-things being added together. Hamilton arrived at them as a spatial generalization of *implicitly* different-things, namely scaled imaginary i plus a real number —complex numbers in short.

Imaginary i was not considered *explicitly* different then, or even now, because it had become considered an ordinary number, misnamed:

> That this subject [imaginary numbers] has hitherto been surrounded by mysterious obscurity, is to be attributed largely to an ill-adapted notation. If, for instance, $+1, -1, \sqrt{-1}$ had been called direct, inverse, and lateral units, instead of positive, negative, and imaginary (or even impossible), such an obscurity would have been out of the question. [Carl Friedrich Gauss in *On Mathematics and Mathematicians*, page 282]

I demur: If, for instance, Gauss had invested even a few minutes studying the foreword of Grassmann's first book after Grassmann sent him a copy (immediately upon getting publisher's copies), he would have realized that imaginaries can be represented by ***directed*** areas "*with the plane of rotation retained*"$_{\text{p}13}$, as Grassmann explained.

Instead Gauss wrote back, in typical Gaussian fashion, that he had already done this a half century before (he had ***not*** retained the plane of rotation), and published it in 1831, and he was very busy. So of course he never read the book.

If he had, he might possibly have realized that *i*'s mysterious obscurity would have been out of the question if +1, −1 had been called positive and negative *ordinary dimensionless units*, and *i* had been called an *extraordinary dimensioned planar unit*.

But that did not happen on this planet. Instead all respectable mathematicians followed Gauss in thinking that a real plus an imaginary was a peculiar "*lateral*" kind of *coalescing same-thing* addition. Almost no one followed Grassmann.

After Hamilton stumbled onto a cohering adamantly *different-thing* addition, he was able to adapt to it, but unable to generalize it. Grassmann, despite a lifetime promoting coalescing same-thing-only addition, was able to adapt to it too after seeing how Hamilton had done it; but he had no time left to generalize it.

Clifford did generalize *cohering different-thing* addition, but then his time expired before he could develop and promote it. And now the idea has still only gained partial appreciation in the mathematics community, and nearly none outside. We humans are very slow about mathematics.

The usual commutative, associative, identity and inverse rules present in the field axioms are not present for the products in these axioms because, with tailored variations, they are *consequences* of the bottom-up rules collaborating with the top-down rules of respect. *Theorems* in short, previously sketched.

Unified semantic axioms

Dimension
Scalars $a, b, \ldots$ have dimension $\{0\}$
0 has dimension $\{\}$
Free units $i_1, i_2, \ldots$ have dimension $\{1\}$
Bound unit $\boldsymbol{i_o}$ has dimension $\{\mathbf{1}\}$
$i_a \ldots i_n$ has dimension $\{\textit{unit-count}\}$
$i_a \ldots \boldsymbol{i_o}$ has dimension $\{\textbf{unit-count}\}$

The dimension of a mixed number $\underline{M}$ is specified by the collective dimensions of the elements in its sum, when distilled minimally.

Metric
The *separation* of a free element $\underline{e}$ is $\sqrt{|\underline{e}^2|}$.
The *magnitude* of bound or free $\underline{e}$ is $\sqrt{|(^{[I]}\underline{e})^2|}$.

The magnitude of a free element vanishes because its free part does. The magnitude of a bound element is the separation of its free part.

This metric, as developed in the syntactic axioms, is orthonormal as presumed by Clifford. To develop a different kind of metric would require abandoning the percolation ploy because then ››ii‹‹ would not always become 1. Its variability would force more-intricate syntactic axioms, whose intricacy could be somewhat accommodated by …

Basis tidiness, part B

The pre-product percolation ploy is an elegant and transparent way to *specify* extension, retraction, extraction; but it is a clunky and inefficient way to actually *compute* them.

For computational efficiency, all bases need to be kept well ordered, and then *zipped together* in ways tailored to each operation. (In my first published version of zipping, I got signs wrong, but my upgraded mind now gets them right in this version, I think.) Tailoring will be imposed on the zipping process, which is common to all operations and always enforces order.

To understand that process, let us begin with extension, whose tailoring is the simplest possible, namely abortion when it encounters a duplicate.

To avoid even that complication at first, let us begin with bases that contain no duplicates; and, for maximum simplicity, require no interleaving, $-2\ i_1i_3i_4$ ◂ $3\ i_5i_7$ for example. Its well-ordered result is obviously $-6\ i_1i_3i_4i_5i_7$. Let us see how the zipping process arrives at that.

The process shall begin at the tail of each base, on its right, because it is useful to immediately encounter the origin if it is present—this ploy is needed for efficient free-part extraction, and also for prompt nonsense detection if retraction is attempted on a bound element.

To clearly visualize the tail-zipping, let us use the raw numeric index of a unit to represent it, and stash the signed-scalar multiplier, -6 in this case, out of sight. Its sign will be constantly modified during the interleaving process.

Under these stipulations the previous extension becomes denoted as 1 3 4 ◂ 5 7. Next, let us embolden the zip-pair "teeth" currently being engaged, 1 3 **4** ◂ 5 **7**, which we shall call *zip-pair* **4 7** Finally, let us display the current $^\text{zipped}$ result in bold superscript on the right and the so-far-unzipped parts in subscript, like so:

$$134 \blacktriangleleft 57$$
$$134 \; 57$$
$$134 \; 5\,^{7}$$
$$134\,^{57}$$
$$\mathbf{13457}$$

Let us walk thru this: In the first step the zip-pair is $_4\,7$. It is already in order, so its right "tooth" becomes the current zipped, namely ^{7}. The next zip-pair is $_4\,_5$, also in order, so zipped becomes $^5\,^7$. There are no zip-pairs left, so the remainder is promoted to zipped, ^{13457}.

Now let us try an extension that requires interleaving, $137 \blacktriangleleft 24$, for example:

$$137 \blacktriangleleft 24$$
$$137 \; 24$$
$$13 \; 24\,^{7}$$
$$13 \; 2\,^{47}$$
$$_{-1} \; 2\,^{347}$$
$$_{-1}\,^{2347}$$
$$\mathbf{-12347}$$

The first zip-pair is $_7\,4$. It is not in order so the $_7$ on the left must be walked all the way across $_2\,_4$ up to the ^{zipped} result. That is two sign changes, effectively none at all. (Of course, efficient computation would not laboriously *walk* units across other units; it would simply note sign change needed to promote a unit to ^{zipped}.)

The next zip-pair is $_3\,_4$, in order, so zipped becomes $^4\,^7$, no sign change. Next is $_3\,_2$, which is not in order so the $_3$ on the left must be walked across $_2$ up to the ^{zipped} result, causing a sign change. Last is $_1$ $_2$ no sign change. Remainder is promoted to zipped, $^{-1}$$2$$3$$4$$7$.

Suppose the extension had contained duplicate units, 1 2 7 ◂ 2 4 say:

$$127 \blacktriangleleft 2\,4$$
$$1\,2\,7\ \ 2\,4$$
$$1\,2\ \ 2\,4\,^{7}$$
$$1\,2\ \ 2\,^{4\,7}$$
$$0$$

This proceeds to zip exactly as before, but when extension notices the duplicate zip-pair $_2$$_2$, it aborts and returns *Nothing*.

Extraction, by contrast, would have returned *Something*, so let us see what that would be:

$$127\ \ \ 2\,4$$
$$1\,2\,7\ \ \ 2\,4$$
$$1\,2\ \ 2\,4\,^{7}$$
$$1\,2\ \ 2\,^{4\,7}$$
$$_1\,^{4\,7}$$
$$\mathbf{1\,4\,7}$$

Notice first that extraction is denoted by mere juxtaposition. It proceeds to zip exactly as extension had done until it encounters the juxtaposed duplicate zip-pair $_2$ $_2$. Its response is that Clifford had decided this should collapse to 1, not 0, which effectively removes that pair from the final result. That result is $^{1\,4\,7}$, meaning of course $i_1 i_4 i_7$ scaled by the hidden signed-scalar.

Notice that this is not an extension, and not a retraction either, good reason to have switched terminology from *extension-retraction* to *extraction*. Incidentally, ¿Does this result appeal to your geometric intuitions in any way? Good luck with that.

Unlike extraction, retraction imposes a rather severe constraint on the zipping process: For it not to vanish, every unit in the short base must be present in the long base. This reflects retraction's geometric meaning as extension undoing: *undoing* requires something already *done*, meaning something entirely within the long base—something *completely parallel* to it.

The basis manifestation of *completely parallel* is *complete duplication of units* amongst the two bases. So let us start with that scenario, and then see what happens when it is not true:

$$2\,4\,7 \bullet 2\,4$$
$$2\,4\,7 \quad 2\,4$$
$$2\,4 \quad 2\,4^{\,7}$$
$$-2 \quad 2^{\,7}$$
$$-7$$

Retraction proceeds to zip exactly as extension and extraction do. Moreover, when it encounters duplicates, it proceeds exactly as extraction does: it collapses them to 1 *after they have been juxtaposed*. That juxtaposition may change signs.

For example, the first duplicate zip-pair that retraction encounters is $_4\,_4$. To get them juxtaposed the $_4$ on the left must be walked across the $_2$. That changes sign, and then that pair vanishes. Then the next duplicate pair, $_2\,_2$ is encountered, but it is already juxtaposed so it vanishes leaving $^{-7}$, meaning $-i_7$ times the hidden signed-scalar.

This is a genuine retraction; but in fact extraction would have computed this product in the same way. So let us look at a retraction that extraction would have computed differently:

$$127 \bullet 2\,4$$
$$1\,2\,7 \quad 2\,4$$
$$1\,2 \quad 2\,4^{\,7}$$
$$0$$

When retraction sees the zip-pair $2\,4$, it knows that the 4 unit in its short base has no duplicate in its long base, so it aborts. ¿How does it know that? Because *it knows that bases are always well ordered.*

So, if a unit in a short base ever encounters a unit in a long base with a smaller numeric index, all hope is gone of ever encountering a duplicate. Hence that unit, i_4 in this case, will end up retracting purely un-duplicated units, all perpendicular to it, so that retraction vanishes.

Extraction would have computed this differently, and in fact already did. The result was $^{-1\,4\,7}$. Extraction never aborts, and always produces something even tho you may not want it, and intend to throw it away.

The duplicate computations in extraction and retraction relied on the normality of Clifford's orthonormal basis. That is not essential, and the 1 previously produced by a duplicate pair can be replaced by a scalar peculiar to that pair. So base-zipping accommodates an arbitrary orthogonal metric.

However, if the basis vectors were not orthogonal, then the various base extensions could not be expressed as pithy Cliffordian products—it would no longer be true that $i_r i_s = i_r \blacktriangleleft i_s$ or that $i_r i_s = -i_s i_r$.

If you understood most of the ideas in this chapter, you have captured the essence of this book. Nice work.

Geometric Algebraic Clarity

In extension theory there appears a characteristic method of calculation which, transposed into geometry, is inexhaustibly fruitful, and here (in the theory of space) consists in subjecting spatial structures (points, lines, and so forth) directly to calculation.

You have seen this before, but I am repeating it here because it is my absolute favorite geometry quote. It comes from the only review of Grassmann's book published in his lifetime, as Grassmann himself stated. He wrote it himself ! at the request of an editor who was baffled by his book. It is best appreciated in alliance with my absolute favorite algebra quote:

One cannot escape the feeling that these mathematical formulas have an independent existence and an intelligence of their own, that they are wiser than we are, wiser even than their discoverers.

Heinrich Hertz wrote that about Maxwell's equations. He was particularly awed by something that had astonished Maxwell: His equations predict electro-magnetic waves that travel at the speed of light. So awed was Hertz about that, in fact, that he diligently searched for such waves. He finally found them by ingenious experiments, which earned his name immortality as *cycles per second*, now denoted as *hz*.

Alliance between Grassmann's geometric composition followed by Hertz's algebraic reckoning is the theme of this chapter—*geometric algebraic clarity*. The procedure is always the same: compose the geometry clearly and let its algebra reckon out the consequences. To begin appreciating the power of this technique, let us start with *fruity algebraic clarity*:

> **Question**: Gwapa Pangit sells fruit. She sold twice as many bananas as mangoes. She sold three more guavas than mangoes. She sells guavas for 15 pesos, mangoes for 5 pesos, and bananas for 10 pesos. She earned 245 pesos during school lunch hour. How many guavas, mangoes, and bananas did Gwapa sell? *Answer*: I want to shoot myself.

Those are not the exact words of Kurt Vonnegut, but they are the essence that I recall (if you have the exact quote, please send it to me). He was a genius with English, and conversant with other languages, but he never learned the foreign language of algebra.

So **he was trying to think it thru**. No no no! Let mathematical formulas do that for you. They are wiser than you, in a mindless way. All that esteemed Kurt Vonnegut needed to do was to *un-English* this into pithy symbols, and then let third-grade arithmetic reckon with it:

- $b = 2m$, $g = m+3$, $15g + 5m + 10b = 245$
- $15(m+3) + 5m + 10(2m) = 245$
- $15m + 5m + 20m = 245 - 45$: $m = 5$, *don't shoot*.
- *Answer*: 5 mangoes, 10 bananas, 8 guavas.

Switching from fruit to geometry, mathematicians have been trying to think it thru since the dawn of history. It's an excellent starter idea, far preferable to not thinking geometry thru; and anyway ¿What else could they do at the beginning? I misspent my youth reading those well-sung …

Think-it-thru Saints

Saint Euclid got *thinking-geometry-thru* streamlined by starting with *"Definitions, Postulates, Common Notions"*, and deriving their consequences. When I tried to understand them as a youth, I discovered they were so tediously presented that maybe only a handful of inhumanly patient geniuses could do so. Try reading *Euclid's Elements* yourself before you even have a girlfriend, or a boyfriend.

(Abe Lincoln got thru the first five books as a self-taught boy. I am in awe. He also got thru the entire King James Bible, where he picked up his English, but seemed not to have picked up Religion with a capital R. Double awe.)

Saint Descartes got *thinking-geometry-thru* efficient computationally by converting everything to scalar *"lengths of certain straight lines"*. His lines, I discovered, were such an awful haystack jumble that, again, I daresay only a few were able to think thru all of them. Try reading *The Geometry* yourself.

(Young Maxwell thought thru much of it, and corrected a mistake. More awe. Incidentally, I was only able to discover one instance of orthonormal so-called "Cartesian" coordinates in this book, and that was a stretch—Newton, I learned, was the first to use them extensively in his undergraduate explorations of cubics, a magisterial feat that he considered unworthy of publishing—*Newtonian coordinates* I call them.)

My next *think-it-thru* Saint, Bertrand Russell, was the only one I could understand as a teenager, and I loved him for it. His *Introduction to Mathematical Philosophy* enchanted me: *Numbers!*—natural, rational, real, complex, cardinal, ordinal … (¿Where are the points?, I would ask now. And also now:

¿Why are the formalities of cardinals and ordinals so utterly useless in science?—what were mathematicians thinking?) His ideas enchanted me too:

I wanted certainty in the kind of way in which people want religious faith. I thought that certainty is more likely to be found in mathematics than elsewhere.

Mathematics may be defined as the subject in which we never know what we are talking about, nor whether what we are saying is true.

Mathematics takes us still further from what is human, into the region of absolute necessity, to which not only the world, but every possible world, must conform.

Mathematics, rightly viewed, possesses not only truth, but supreme beauty—a beauty cold and austere, like that of sculpture, without appeal to any part of our weaker nature, without the gorgeous trappings of painting or music, yet sublimely pure, and capable of a stern perfection such as only the greatest art can show

The fact that all Mathematics is Symbolic Logic is one of the greatest discoveries of our age; and when this fact has been established, the remainder of the principles of mathematics consists of the analysis of Symbolic Logic itself.

The pure mathematician, like the musician, is a free creator of his world of ordered beauty.

To a mind of sufficient intellectual power, the whole of mathematics would appear trivial, as trivial as the statement that a four-footed animal is an animal.

The human race may well become extinct before the end of the century. Speaking as a mathematician, I should say the odds are about three to one against survival.

As a young naif I did not see incongruity in Russell's ideas; but as an old sophisticate I do. First of course, here we are surviving well after his century. So clearly, when speaking as a pure mathematician, Russell was not computing as an

304

impure statistician.

Second, ¿If a mathematician truly is a creator of his world, how could mathematics ever appear trivial and absolutely necessary?—creations are never necessary, and good ones are never trivial.

To be fair, Russell eventually recognized and acknowledged the flaws in his thinking, which makes him a true hero in my mind—few people ever do that. He asserted that he loved to change his ideas.

Unfortunately, as a tired oldster he did not publicly flaunt the changes he had made to ideas he had flaunted as an eager youth. Instead he acknowledged them privately, for example at a party hosted by his mathematics friend Littlewood, who wrote ...

> I felt the theory [General Relativity] was about the greatest intellectual advance and illumination that ever happened. I explained it to Russell, who at that time knew no physics. He was similarly staggered. Suddenly he burst out (to Dora's consternation): *"To think I have spent my life on absolute muck."* [*Littlewood's Miscellany*, page 129]

The muck he was talking about, we know from his later letters, was the idea that *"all mathematics is Symbolic Logic"*, as presented in his *Principia Mathematica*. As an old man I agree—absolute muck: mathematics truly is a free creation of humans—not muck at all. Here is my mature postjudice of his muckiness:

- Certainty in mathematics—not entirely muck.
- We never know what we are talking about nor if what we are saying is true—absolute muck, vacuous witticism.
- Mathematics takes us from what is human to absolute

necessity—half muck: it takes us from basely human to divinely human, by no means necessary.

- Bold austere beauty—slight muck: bold beauty, not austere.

- The human race may not survive—not muck at all; we are fast becoming different creatures, better creatures possibly; and our planet is developing enough online immunity against global war to let us do so. My hope is that we will become divinely un-human, meaning all conversant in geometric algebra as a native tongue. Ha.

- Remember, these are only personal opinions in a final chapter—I am often wrong, as Russell asserted he often was.

Since this section is about *thinking geometry thru*, here is a sampler of how Russell did that, presented in *An essay on the Foundations of Geometry*. It was "based on a dissertation submitted at the Fellowship Examination of Trinity College, Cambridge, in the year 1895." You can decide for yourself about its relative degrees of muck.

At the same time, it is a pity that Riemann, in accordance with the metrical bias of his time, regarded space as primarily a magnitude, or assemblage of magnitudes, in which the main problem consists in assigning quantities to the different elements or points, without regard to the qualitative nature of the quantities assigned.

That the notion of imaginary points is of supreme importance in Geometry, will be seen by any one who reflects that the circular points are imaginary.

The metaphysician who should invent anything so preposterous as the circular points would be hooted from the field. But the mathematician may steal the horse with impunity.

Now a relation between two points can only be defined by a line joining them—nay further, it may be contended that a relation can only be a line joining them.

Now externality to the Self, it would seem, must necessarily raise the whole question of the nature and limits of the Ego, and what is more, it cannot be derived from spatial presentation, unless we give the Self a definite position in space.

The essence of my contention is that, if experience is to be possible, every sensational This must, when attended to, be found, on the one hand, resolvable into Thises, and on the other hand dependent, for some of its adjectives, on external reference.

The antinomy of the point—which arises wherever a continuum is given, and elements have to be sought in it—is fundamental to Geometry. It has been given, perhaps unintentionally, by Veronese as the first axiom, in the form: "There are different points. All points are identical."

The pure doctrine of extension, as constructed by Grassmann, need not be discussed—it included much empirical material, and was philosophically a failure.

[And so on and on and on.]

The doctrine of extension, as constructed by Grassmann, was used by Russell's *Principia* co-author, Alfred North Whitehead, to essentially package almost everything he had ever learned about mathematics so far, beginning in third grade. It was entitled *A Treatise on Universal Algebra* (these two loved grandiose titles), published in 1898.

Whitehead had guided Russell's exposition of his *Essay* four years earlier, as Russell acknowledged; but Russell seems to never have had the slightest interest in Whitehead's subsequent book, or knowledge of its "*empirical*" kind of geometry. I chalk that up to his pompous *think-it-thru* personality.

Which cost him, in his own mature assessment, a lifetime of muck in mathematics. Beyond mathematics "*He was the most fascinating man I have ever known, the only man I ever loved, the greatest man I shall ever meet, the wittiest, the gayest, the most charming. It was a privilege to know him, and I thank God he was my father.*" Lady Katherine Tait.

Geometric Language Creators

These are the true heroes in my book. They offloaded the *thinking-it-thru* part onto algebra. All you have to do is *un–prose* the geometry—express it in pithy symbols—and then let the algebra reckon with it.

Their patron Saint, of course, is Grassmann; but those few visionaries who embraced his new language had to re-create it in their own ways because, frankly, few could understand Grassmann, and he somewhat botched the job, in my view, as first-creators usually do.

The most-diligent re-creator was the one just mentioned, Alfred North Whitehead. He was one of the few: he regarded Grassmann's abstract and esoteric presentation as a fantastic philosophical success, in total contrast to Russell:

> The greatness of my obligations in this volume to Grassmann will be understood by those who have mastered his two *Ausdehnungslehres*. The technical development of the subject is inspired chiefly by his work of 1862, but the underlying ideas follow the work of 1844. At the same time I have tried to extend his Calculus of Extension both in its technique and in its ideas.px

He launches his exposition by stating that "*Universal Algebra is the name applied to that calculus which symbolizes general operations, defined later, which are called Addition and Multiplication.*"p18

That seems refreshingly simple, but these are not your

308

teenager's Addition and Multiplication (sadly)—they envelop, in Whitehead's presentation *"Descriptive Geometry, Line Geometry, and Metrical Geometry, both non-Euclidean and Euclidean."*_{pix}

And also Riemann Manifolds, Symbolic Logic, Quadratics, Intensity, Combinatorial Multiplication, Regressive Multiplication, Progressive Multiplication, Supplements, Conics and Cubics, Matrices, Groups, Elliptic Geometry, Hyperbolic Geometry, Kinematics, Curves and Surfaces, Parabolic Geometry, and finally beginning on page 505, Extension.

Whitehead had been flying high before *Extension*, but now he descends to earth and begins to present a coherent sequence of figures. However, if you haven't seen this before, you likely won't understand it here, even tho Whitehead obviously did.

One of the ideas he understood clearly is something that has been lost to the current geometric-algebra community, namely the distinction between free and bound.

He called free elements of all dimensions *vectors*, meaning *carriers*; and he called free-part extraction *"the operation of taking the vector"*._{p516} He ended his book in the purely free sub-algebra, which he called *Pure Vector Formulae*, *"dropping altogether the representation of the point as a primary element."*_{p548}

Prior to that, ***points had been numbers*** for Whitehead, ***primary*** numbers, able to participate in Addition and Multiplication. Points never were numbers for his co-author Russell or any of the other *think-it-thru* Saints right down to the present day, Thursday, September 12, 2019.

In conclusion, Whitehead writes "*It must be distinctly understood that the present work does not pretend to exhaust the ideas in Grassmann's two versions of the 'Ausdehnungslehre': I only deal with those parts, which I have been able to develope and to bring under one dominant idea.*"p573

¿What dominant idea? Addition and Multiplication of "*points, lines and so forth*", Grassmann's "*characteristic method of calculation*" that this chapter opened with.

The other one able to re-create Grassmann's language in the 1800's was Giuseppe Peano. This is nearly unknown, even among mathematicians, who associate him almost exclusively with *Peano's axioms*, the rules that recursively generate natural numbers by endlessly adding 1.

If history had been more respectful, those rules would now be called *Grassmann's Induction* because they were first presented in a book on arithmetic that Grassmann wrote for his teaching duties. Peano discovered them in that book (testimony to his insatiable curiosity), as he acknowledged, and then presented them in his own tidy way.

Peano tried to make history respectful by also presenting Grassmann's Extension Theory in his own tidy way, titled *Geometric Calculus*, published in 1888. He had high hopes for it:

> The geometric calculus exhibits analogies with analytic geometry; but it differs from it in that, whereas in analytic geometry the calculations are made on numbers that determine the geometric entities [Cartesian coordinates], in this new science the *calculations are made on the geometric entities themselves* [my emphasis] …
>
> … [Grassmann's algebra] to a great extent incorporates the others and is superior in its powers of calculation and in the simplicity of

its formulas. But the excessively lofty and abstruse contents of the *Ausdehnungslehre* impeded diffusion of that science ...

It is however my opinion [in the 1880's] that before long this geometric calculus, or something analogous, will be substituted for the methods actually in use in higher education. It is indeed true that the study of this calculus, as with that of every other science, requires time; but I do not believe that it exceeds that necessary for the study of, e.g. the fundamentals of analytic geometry; and then the student will find himself in possession of a method which comprehends analytic geometry as a particular case, but which is much more powerful ...

¿*Before long*? Going on one hundred and thirty years now? The study of Peano's *Calculus* required a pleasant week of my time. It gave me hope for us humans, that we are curious enough and diligent enough to create as Peano had created.

His presentation is more tidy than Grassmann's; and more focused, comprehensible, and pithy than Whitehead's (144 pages vs 573). It is essentially a word-picture book without pictures; so your best hope of understanding it is to provide the pictures yourself. I like doing that whenever an author is graphic enough to enable it, as Peano is, so studying his book was fun.

Nonetheless, it was like visiting a foreign country whose language and customs are unfamiliar: bound vectors, bivectors, trivectors are called *lines, surfaces, volumes*. To get their free parts, you **construct** their *vector, bivector, trivector* parts, respectively.

That language was easy enough to adapt to; but I had trouble with the customs: In Peano country you don't *calculate* things like you do in Whitehead country, rather you *"construct"* things, or *"produce"* them, assisted by a looming helper.

311

You have no choice about that because everything in the language is defined geometrically in terms of *"formations"* relative to a *"unit volume"*, and it is really strange—it is a tetrahedron that follows you everywhere.

Peano has assigned it as your guide for all your interactions with the natives. Its first job, after you check in to your hotel, is to demonstrate how it gets negated when you exchange its vertices. (I was perplexed by this abrupt display—¿Shouldn't we be starting simpler?)

Then this tetrahedron shows you that its interactions with the lessor natives makes the ordering of their terms irrelevant, makes them equal to each other when they equal the same thing, gives them all bondage to the space thru themselves; and it generally lets you know that all your interactions with those lessors will be mediated by It.

It will be doing the constructing and the producing, not Addition and Multiplication, which serve merely to relate everything to the tetrahedron, who will be doing the heavy lifting. And by the way, it will be sleeping with you.

So I am never going back to that tidy little country. But I am glad to have visited because its fidelity to Grassmann's conceptions clarified for me precisely how Grassmann had somewhat botched the job as far as I am concerned:

In his mature algebra, most interactions are mediated by the Ceiling of your ambient space, whatever that may be. Things are not allowed to have personal relations between themselves —the Ceiling has to get involved.

Geometric composition

Finally we come to the central problem in this chapter:

¿How do we best compose our geometry so that the algebra can reckon with it effectively? The extraordinary expressiveness of Grassmann-Clifford algebra makes this essentially an *un-prosing* problem; but a more intricate one than un-Englishing a word problem.

The guiding principle is to use encompassing *extraction* as much as possible. Its versatile symmetries—associativity, identity, element inverses—make it easy to compose and easy to reckon with; and the way it fully informs about geometric relationships makes it pithy and concise.

Our current geometric-algebra community does this reflexively by having become accidentally trapped in the free sub-algebra, something that truly amazes me.

I believe it happened like this: Altho the entire Grassmann-algebra community in the 1800's was intimately familiar with bound and free elements, that community was never larger than a handful of cognoscenti: Clifford, Peano, Whitehead and maybe half a dozen less smitten.

After Gibbs' new (de-facto free) vector algebra became wildly popular beginning in the early 1900's, the Grassmann community just died off, to put it bluntly. That community had left a record of its existence via the expositions just scrutinized, but they were as unreadable as the *Principia* was, even to specialists.

Anyone who had trouble reading Grassmann's *Ausdehnungslehre*s—meaning virtually everyone who ever tried except possibly Clifford—would have found Whitehead's packaging of them with *all-known-mathematics-at-that-time* even less readable. Anyone who tried to comprehend Peano's condensation of them around a tetrahedron would have been baffled. Clifford's published short enthusiasms about Grassmann were about the only understandable glimpses later mathematicians had.

The only comprehensive modern treatment of *Grassmann Algebra* that I know of is John Browne's exposition with that title. This may be the clearest exposition of it ever written, and it has the further advantage of being computable, but only within Wolfram's *Mathematica* behemoth.

However, it still requires you to sleep with the Ceiling. It is a nicer companion here—a parallelepiped rather than a tetrahedron—which gives it better scaling relations between its bound and free manifestations: direct correspondence rather than a factorial.

(Peano had hidden the factorial relation by cavalierly *"we-say"*ing that a free parallelepiped has the same magnitude as a corresponding bound tetrahedron. That is when I left the country.)

If you must be trapped in an algebra, then an extremely expressive one like the Cliffordian free sub-algebra is ideal. However there are times when you really *really* need points to, say, find the center of mass, or compute a central force. And there are times when you really need bound vectors to properly locate a force or a momentum. And so on. The literature exhibits two common responses to their absence.

The first is to avoid ever getting in those situations. This is the *search-under-the-streetlight* tactic because that is where the illumination is; even tho you know you dropped your keys in a dark alley.

The second response, to switch metaphors, is to use your all-purpose handyman tool in ways for which it is ill suited:

314

Open up the screwdriver blade, for example, to pry a point into place. Or bang the tool sideways against a bound vector to drive it home.

So-called *position vectors* are usually involved in these shady activities, but that term is an oxymoron because these vectors are as free as everything else in the free sub-algebra.

The full geometric algebra allows you to select precisely the right tool for the job; and you can easily move between a bound tool and free tool like so: To bind something free, simply extend it with a point wherever you want it located. To free something bound, just extract its free part.

Unfortunately, using the expressiveness of the full algebra opens up the possibility that you may generate geometric nonsense by retracting or extracting something bound. The early Grassmann kept trying that with retraction. The mature Grassmann stopped, but evidently felt you should have the same opportunity he had—he left it up to you to avoid nonsense.

So the question arises: ¿Should we prohibit retraction \ extraction of bound things like we prohibit division by zero? The answer is that they generate different kinds of nonsense —the former is *semantic* nonsense, the latter is *syntactic* nonsense.

Syntactic nonsense is a non-starter in any language; but semantic nonsense is always available in any adequately informative language. Historically, the only effective response has been free speech—language policing has never worked. Here is Hamilton's characteristically Hamiltonian advocacy of that:

Mr Jarrett brought forward again his Cambridge objections [to

315

quaternions], and dwelt particularly on the possibility of making mistakes in the use of my new calculus. In reply to which I disclaimed the power of setting any limit on the faculty of making blunders …

Mr. Airy, seeing that the subject could not be cushioned, rose then to speak of his own acquaintance with it, which he avowed to be none at all; but gave us to understand that what he did not know could not be worth knowing.

[*A History of Vector Analysis*, page 34]

If, in articulating points, lines and so forth, we can't set any limit on the faculty of making blunders, perhaps the algebra we hand it off to can? Maybe some kind of spell-checking or grammar-checking?

Algebraic reckoning

Nowadays algebra is done using an algebra program on a computer. For such a program, grammar-checking is not an option that can be turned off, like it can be in a word processor—your grammar *must* be impeccable before the program will even consider reckoning with it.

If the program finds your geometric prose acceptably literate, it begins trundling thru it by using respect for coherence to descend all the way down to bases. When it arrives at base products—extension, retraction, extraction—it would, for efficiency, compute them by zipping them together in tailored ways, as described in the previous chapter.

This process is fraught with peril, and here is where your composition, tho syntactically impeccable, can easily slip off the rails. To illustrate, here are some of the possibilities in JavaScript: undefined, NaN (not-a-number), null, stop-the-train!, meaning an exception that halts execution.

Your composition is susceptible to all of these problems because the algebra program itself is. It is good to be so-susceptible because it helps avoid computational nonsense.

It would be even better to be even *more* susceptible to help avoid *geometric* nonsense. For example, the algebra program could easily notice, before it begins to retract two bases, whether either base is bound—it would see the origin i_o immediately.

¿When it sees it, should it stop the train? Return NaN? Undefined? null? Or should it just ignore the faux pas and keep that train rolling, even tho it will probably generate nonsense?

Here is the conundrum: it may not generate nonsense. For example, it is perfectly meaningful to retract by the origin. Consider this retraction: $i_1 i_2 i_o \bullet i_o$. It generates $i_1 i_2$, the free part of $i_1 i_2 i_o$. Now consider this retraction: $i_1 i_2 \bullet i_o$. It generates 0, the free part of $i_1 i_2$.

In other words, retraction by the origin automatically generates both of the bottom-up rules for free-part extraction. There was no need to spell those rules out—that operation could have been succinctly defined as retraction by the origin.

¿So why didn't I define it that way? Why did I explicitly spell out those two rules if there was no need to? I felt there was a need to because this kind of bound retraction is about the *only* one that does not generate nonsense.

Defining it separately, explicitly, allows the algebra program to uniformly inform about the other bound retractions that do generate nonsense. There are several informing options. The most obvious would be to just stop the train. I suspect most geometric composers would find that unacceptable because they would have to keep jumping off and back on.

A kinder, gentler approach would be for the algebra to return a formal *nonsense* result when a bound retraction is encountered. The train would keep rolling, but all subsequent calculations would propagate that result, so the final result would be *nonsense*.

When that result is seen, a composer could stay in her seat, look back to find out where the *nonsense* started, recompose, and get the train rolling again.

That is how *undefined* works in JavaScript, but it might be unwise to borrow that language-generic value, or even *NaN*, for use in the algebra program—it needs a specific *geometric* kind of undefined or not-a-number.

Even the *nonsense* approach might not be gentle enough for extraction, which might be intended by a composer as a convenient extension, whose retraction part is being ignored or filtered out. The algebra would have no way of knowing that, so what is it to do?

The algebra does know that the retraction part is very likely nonsense, so it could just keep on reckoning and annotate the final result with a message like this: *possible retraction nonsense*. The most sensible response to such a message would be to switch to explicit extension.

Possibly the best algebraic strategy is to give the geometric composer all of these options, and allow her to select the ones she likes, or none at all.

Esteemed Reader, I am delighted you made it all the way to the end, even if you got here by flying low; and I feel compelled to make the very last words you encounter pregnant with meaning, fraught with wisdom, pleasantly pompous, but encapsulating nonetheless the core message in this chapter, in this book:

Semantics engenders Syntax

^
.+.
^

but syntax engenders a cat.

www.ingramcontent.com/pod-product-compliance
Lightning Source LLC
Chambersburg PA
CBHW080954170526
45158CB00010B/2801